LRFD Bridge Design

LRFD Bridge Design

Fundamentals and Applications

Tim Huff

CRC Press

Taylor & Francis Group

Boca Raton London New York

CRC Press is an imprint of the
Taylor & Francis Group, an **Informa** business

First edition published 2022
by CRC Press
6000 Broken Sound Parkway NW, Suite 300, Boca Raton, FL 33487-2742

and by CRC Press
2 Park Square, Milton Park, Abingdon, Oxon, OX14 4RN

Library of Congress Cataloging-in-Publication Data

Names: Huff, Tim, author.
Title: LRFD bridge design : fundamentals and applications / Tim Huff,
Tennessee Technological University Cookeville, Tennessee.
Description: First edition. | Boca Raton : CRC Press, [2022] | Includes
bibliographical references and index.
Identifiers: LCCN 2021043502 (print) | LCCN 2021043503 (ebook) | ISBN
9781032208367 (hardback) | ISBN 9781032208374 (paperback) | ISBN
9781003265467 (ebook)
Subjects: LCSH: Bridges--Design and construction--Textbooks. | Load factor
design--Textbooks.
Classification: LCC TG300 .H84 2022 (print) | LCC TG300 (ebook) | DDC
624.2/5--dc23
LC record available at https://lccn.loc.gov/2021043502
LC ebook record available at https://lccn.loc.gov/2021043503

ISBN: 978-1-032-20836-7 (hbk)
ISBN: 978-1-032-20837-4 (pbk)
ISBN: 978-1-003-26546-7 (ebk)

DOI: 10.1201/9781003265467

Typeset in Times
by Deanta Global Publishing Services, Chennai, India

Access the Support Material at www.routledge.com/9781032208367

Contents

Preface

Beginning with basic concepts in bridge geometry, the text progresses through discussions on the various elements of typical I-girder bridges. Criteria from the 9th edition of the American Association of State Highway and Transportation Officials (AASHTO) LRFD Bridge Design Specifications are presented and applied to sample problems. Many examples are based on constructed bridges designed by the author. Steel and concrete I-girder design, deck and parapet design, load calculations, bearing design, and substructure design are all included. The book ends with chapters devoted specifically to seismic design and seismic isolation applied to bridges. Each chapter ends with a section of solved problems to illustrate the principles covered, followed by exercises which may be used as exam problems by instructors. An Appendix consists of detailed solutions for the exercises.

This book is intended for use as a reference for practicing bridge engineers and as a textbook for a course (or multiple courses) in bridge engineering.

Tim Huff is a faculty member of the Civil & Environmental Engineering Department at Tennessee Technological University in Cookeville, where he resides with his beautiful and talented wife, Monica, an artist and a teacher.

This book is dedicated to my family – Monica, Majo, Esteban, Troy, Holli, and my parents, Bill and Sue. My inspiration for the book is my students at Tennessee State University and Tennessee Tech University, as well as the experiences encountered over the course of 35 years as a practicing structural engineer.

Blessings and peace to all, without qualification. May you be happy and healthy, free from pain and suffering, and may you find joy and peace always.

Acknowledgements

I thank my mentors for their confidence in me, my co-workers for their friendship, my students for teaching me more than I taught them. I thank my family for love and inspiration.

The Bridge Builder

An old man going a lone highway,
Came, at the evening cold and gray,
To a chasm vast and deep and wide,
Through which was flowing a sullen tide.
The old man crossed in the twilight dim;
The sullen stream had no fear for him;
But he turned when safe on the other side
And built a bridge to span the tide.

"Old man," said a fellow pilgrim near,
"You are wasting your strength with building here;
Your journey will end with the ending day;
You never again must pass this way;
You have crossed the chasm, deep and wide,
Why build this bridge at eventide?"

The builder lifted his old gray head:
"Good friend, in the path I have come," he said,
"There followeth after me today
A youth whose feet must pass this way.
This chasm that has been as naught to me
To that fair-haired youth may a pit-fall be.
He, too, must cross in the twilight dim;
Good friend, I am building the bridge for him."

Will Allen Dromgoole

Author biography

Tim Huff has 35 years of experience as a practicing structural engineer. Dr Huff has worked on building and bridge projects in the United States and has contributed to projects in India, Ethiopia, Brazil, the Philippines, and Haiti as a volunteer structural engineer with Engineering Ministries International. He is a faculty member of the Civil & Environmental Engineering Department at Tennessee Technological University in Cookeville, where he resides with his beautiful and talented wife, Monica, an artist and teacher.

1 Introduction

This course is intended for senior level undergraduate civil engineering majors with an emphasis on structural engineering, and graduate level structural engineering students. Practitioners may also find the material to be a valuable reference. After this introductory chapter (Chapter 1) on preliminary design considerations and construction stages in the life of a bridge, subsequent topics covered include the following:

- load calculations and limit states: Chapters 2 and 3
- bridge deck design: Chapter 4
- live load distribution: Chapter 5
- steel welded plate girders: Chapter 6
- prestressed concrete girder bridges: Chapter 7
- bridge girder bearings: Chapter 8
- reinforced concrete substructures and foundation design: Chapter 9
- seismic design concepts for bridges: Chapter 10
- seismic isolation applied to bridges: Chapter 11

Load and resistance factor design (LRFD) of I-girder-type bridges is the basis of the superstructure-related content in Chapters 5 through 8. Primary references for bridge design include specifications from the American Association of State Highway and Transportation Officials (AASHTO). These documents, along with the shorthand notation used for each in this book, are summarized below.

- AASHTO LRFD Bridge Design Specifications, 9th edition (AASHTO, 2020). Hereafter referred to as the LRFD BDS.
- AASHTO Guide Specification for LRFD Seismic Bridge Design, 2nd edition (AASHTO, 2011). Hereafter referred to as the LRFD GS.
- AASHTO Guide Specification for Seismic Isolation Design, 4th edition (AASHTO, 2014). Hereafter referred to as the GS ISO.
- AASHTO LRFD Bridge Construction Specifications, 4th edition (AASHTO, 2017). Hereafter referred to as the LRFD BCS.

1.1 THE PROJECT BRIDGE

The Project Bridge, defined by Figures 1.1 through 1.7, provides the basis for much of the discussion on the design of the various elements of a typical I-girder bridge. Some of the examples in each chapter are based on this Project Bridge. Consisting of two 90-ft spans, the 34-ft 5-inch- wide Project Bridge design will include both prestressed concrete I-girder and steel I-girder superstructure options. Deck design,

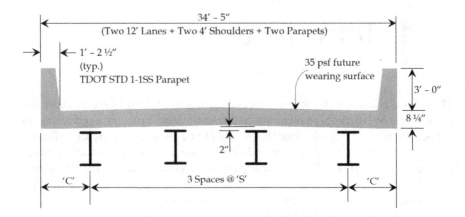

FIGURE 1.1   Bridge cross section.

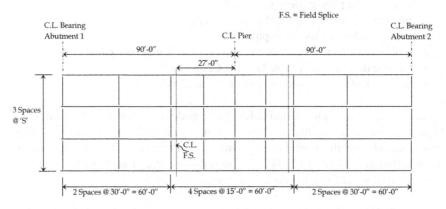

FIGURE 1.2   Bridge framing plan (steel girder option).

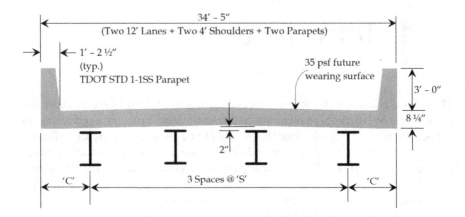

FIGURE 1.3   Plan view – intermediate diaphragms (concrete girder option).

stability bracing (for the steel option), bearing design, bent (pier) design, and foundation design will all be explored.

Figures 1.3 through 1.6 for the concrete girder option were generated using the *LEAP Bridge Concrete* software by Bentley. With prestressed concrete I-girder bridges, intermediate diaphragms, shown at the one-third span points in Figure 1.3,

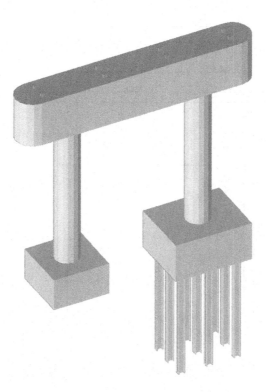

FIGURE 1.4   Isometric of pier.

are often provided to improve stability during construction. It is not uncommon for these diaphragms to have "standard" details with little or no design provisions. Precast, prestressed I-girders typically span from support to support and behave as simply supported beams until the cast-in-place deck attains design strength.

Figure 1.2 for the steel I-girder option depicts cross-frames between the gird-ers spaced 30 ft apart in the positive moment (compression in the top of the deck) region and at 15 ft in the negative moment region. These cross-frames become critical elements for stability, particularly during construction, for long span steel I-girder bridges. Careful attention to design requirements becomes necessary in such cases. Figure 1.7 for the steel I-girder alternate identifies field splice locations. Steel I-girders constructed in such a manner behave as continuous beams for all loads, though part of the load is carried by the non-composite section and part by the com-posite section. More discussion on continuity will be reserved for later. Each field section is limited by shipping and handling requirements and varies from place to place, and among fabricators and contractors.

A brief discussion on the general behavior of each major element is in order, prior to developing detailed design requirements.

First, consider the parapet in Figure 1.1. This element behaves much as a can-tilever beam subjected to primarily lateral loads. Certainly, wind loads act on the

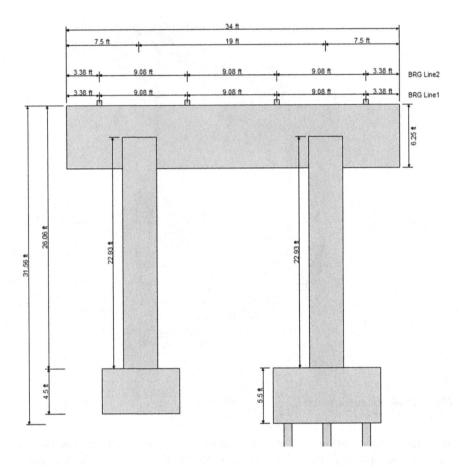

FIGURE 1.5 Basic pier dimensions (preliminary).

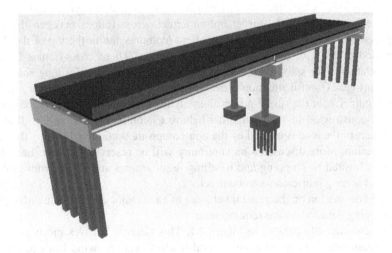

FIGURE 1.6 Isometric of entire bridge.

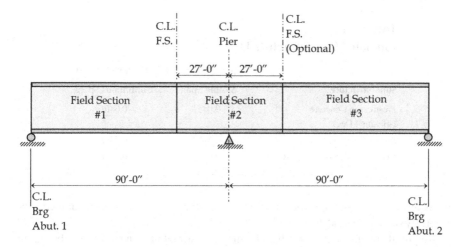

FIGURE 1.7 Elevation view of bridge (steel girder option).

parapet. But the primary loading affecting parapet details is a vehicular collision load, typically assumed to be a lateral load applied at the top of the cantilever parapet. The load for which a parapet must be designed will hopefully never occur. Second, consider the typically cast-in-place concrete deck, which behaves as a continuous beam between I-girders. The deck will experience tension in the top fiber (negative moment) over the supports (the I-girders) and tension in the bottom fiber (positive moment) midway between the supports. The negative moment condition requires consideration of two cases: (1) the interior negative moment over a support and (2) the overhang negative moment produced by the parapet weight, the deck self-weight, any overlay, and any traffic outside the exterior girder. The I-girders themselves will also behave as continuous beams (at least for part of the load) spanning between adjacent abutments and piers. Whereas these I-girders could be designed using a complex two-dimensional grillage model, more often the design is completed with a line-girder analysis of a single girder. This concept will be explored in later sections, primarily in Chapter 5 on live load distribution. Finally, consider the bent (pier) in Figure 1.4. This is typically a reinforced concrete rigid frame in the transverse direction. Behavior in the longitudinal direction may be either as a rigid frame or as a cantilever.

1.2 PRELIMINARY DIMENSIONS

Before design of the various bridge elements may commence, it is necessary to establish the cross-section geometry in terms of girder spacing, 'S', and overhang (cantilever) dimension, 'C' (see Figure 1.1). Although it is theoretically acceptable to use larger overhangs, it is usually most economical, in terms of balancing deck moments and interior/exterior girder live load distribution, to use a value for 'C' of no more than 40% of 'S'.

Though not a requirement, it seems most advantageous to provide a girder spacing of 8 ft to 10 ft for prestressed girders, and 9 ft to 12 ft for steel I-girders, based

TABLE 1.1

Estimated Superstructure Depth

Superstructure	Approximate Depth of Superstructure	
	Simple Spans	Continuous Spans
Prestressed Concrete I-Beams + Deck	$0.045L$	$0.040L$
Steel I-Girder + Deck	$0.040L$	$0.032L$
I-Girder Portion of Steel I-Girder	$0.033L$	$0.027L$

on experience. Certainly, girder spacing outside these ranges has been adopted successfully on many projects.

Required superstructure depths, D_{ss}, may be estimated using Table 1.1, based on Section 2.5 in the AASHTO LRFD BDS. In the tabulated expressions, 'L' is the span length. The table is a rough guide, not a strict requirement, and may be used to establish preliminary superstructure depths. Often, the final design depth will not satisfy the tabular estimates.

The estimates are traditionally recommended (as opposed to required) minimum values. Nonetheless, they may prove valuable in establishing a starting point for required girder depth. Later, for both the concrete and steel girder options, final selected dimensions will be larger for the Project Bridge than these computed, recommended, minimum values. The primary consideration in superstructure depth is vertical clearance. Bridges over highways typically require a minimum vertical clearance of 16 ft from the roadway below to the bottom of the girder. To allow for future overlay of the road below, it is common practice to provide 16 ft 6 inches of vertical clearance (see the LRFD BDS, Section 2.3.3.2). For river and stream crossings, vertical clearance is a design consideration for the hydraulic properties of the structure and the low girder elevation. The vertical clearance requirement for bridges is 23 ft, as governed by the Manual for Railway Engineering (AREMA, 2021).

Horizontal clearance is important in the placement of substructures. A 30-ft clear zone is typically adopted for high-volume roadways. The clear zone is the "unobstructed, traversable area provided beyond the edge of the through-traveled way for the recovery of errant vehicles" (AASHTO, 2011). The clear zone is generally measured from the edge of the outermost lanes, though there are exceptions. Horizontal clearance is measured perpendicular to the roadway below. Skewed bridges thus require longer spans to satisfy horizontal clearance requirements. Navigable waterways often require very long spans to accommodate horizontal clearance requirements for water traffic in the form of barges and ships.

Terminology for intermediate supports varies. End supports for bridges are called abutments. Intermediate supports are called either piers or bents. One frequently used distinction is to use the term "pier" for water crossings and "bent" for roadway and railway crossings. The two terms, pier and bent, will be used interchangeably in this book.

1.3 BRIDGE GIRDER BEHAVIOR AT VARIOUS STAGES OF CONSTRUCTION

Although there are exceptions, prestressed girder bridges are typically constructed so as to behave as follows:

- non-continuous and non-composite for self-weight, deck weight, intermediate diaphragms, and construction live loads
- continuous and composite for parapets, sidewalks, future overlay, and traffic

Once again, there are exceptions, but steel I-girder bridges, on the other hand, are typically designed and constructed so as to behave:

- continuous and non-composite for self-weight, deck weight, cross frames, lateral bracing, and construction live loads
- continuous and long-term composite for parapet, sidewalks, future overlay, and utilities
- continuous and short-term composite for traffic

Long-term composite properties for steel I-girders are computed using steel as the primary material with a modular ratio of $3n$ applied to concrete. Short-term composite properties for steel I-girders are computed using a modular ratio, $n = E_S/E_C$ (ratio of steel to concrete Young's modulus). The long-term property calculations are intended to account for creep under permanent loads, while research has shown that short-term properties are more appropriate for transient loads.

For prestressed girders, continuity is achieved by extending strands beyond the beam end at the pier, bending the strands up, and embedding the strands into a cast-in-place diaphragm. These bent-up strands are not shown in Figure 1.8 and will be discussed in later sections. Prestressed concrete girder continuity also requires

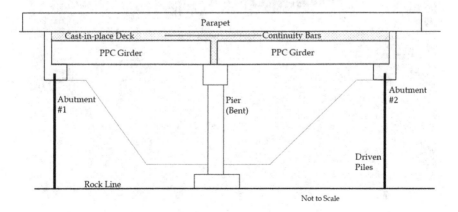

FIGURE 1.8 Schematic of a typical prestressed concrete girder bridge.

relatively heavy longitudinal reinforcement in the deck to resist negative moments at interior supports.

Figures 1.8 and 1.9 are schematic depictions of typical prestressed concrete and steel I-girder bridges, respectively. Figures 1.10 and 1.11 are photographs of actual structures under construction.

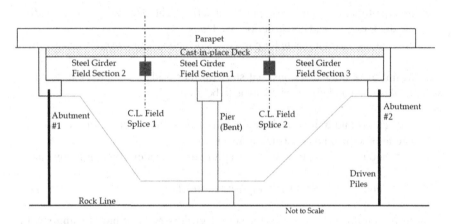

FIGURE 1.9   Schematic of a typical steel I-beam bridge.

FIGURE 1.10   Demonbreun Street Viaduct in Nashville, TN, under construction.

FIGURE 1.11 SR-70 over Center Hill Lake in Smithville, TN, under construction.

1.4 BRIDGE MATERIALS

Properties for some of the reinforcing steels commonly used in bridge construction are summarized in Table 1.2. A 706 reinforcement has a higher reliable ultimate strain than does A 615 reinforcement, and has a maximum cap on yield strength, both characteristics being beneficial in the seismic design of ductile elements.

A 706 reinforcement is required in plastic hinging regions for longitudinal reinforcement of columns in regions of high seismic hazard (see Section 8.4.1 of the LRFD GS). For computing overstrength plastic moment resistance in seismic analysis, the specified magnifier (λ_{mo}) is equal to 1.2 for A 706 bars and 1.4 for A 615 bars (see Section 8.5 of the LRFD GS).

Table 1.3 provides design information for prestressing steels used in bridge construction. Low-relaxation ("low-lax") strands generally experience less loss of prestress force compared to their stress-relieved counterparts and are a popular selection for modern bridges. Properties of two of the most frequently encountered 270K, 7-wire strands, whether low-lax or stress-relieved, are:

- ½-inch diameter strands: $A = 0.153$ in²/strand
- 0.6-inch diameter strands: $A = 0.217$ in²/strand

Standard AASHTO I-beam and Bulb-T girders are depicted in Figures 1.12 and 1.13, respectively, with properties listed in Tables 1.4 through 1.6. The figures are

TABLE 1.2
Reinforcing Steels

ASTM	Grade	f_y, ksi	f_u, ksi	f_{ye}, ksi	ε_{cl}	ε_{tl}	ε_{SU}	ε_{SU}^R
A 615	40	40	60	68				
	60	60	90		0.0020	0.0050	0.09 #4-#10 0.06 #11-#18	0.06 #4-#10 0.04 #11-#18
	75	75	100		0.0028	0.0050		
	80	80	105		0.0030	0.0056		
A 706	60	60 min 78 max	80	68	0.0020	0.0050	0.12 #4-#10 0.09 #11-#18	0.09 #4-#10 0.06 #11-#18
	80	80 min 98 max	100		0.0030	0.0056		
A 1035	100	100	150		0.0040	0.0080		

TABLE 1.3
Prestressing Steels

Material	Diameter	f_{pu}, ksi	f_{py}, ksi	E, ksi
Low-relaxation 7-wire strand	0.375–0.600	270	$0.90f_{pu}$	28,500
Plain bars	0.750–1.375	150	$0.85f_{pu}$	30,000
Deformed bars	0.625–2.500	150	$0.80f_{pu}$	30,000

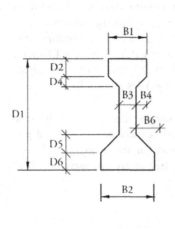

Type I-IV

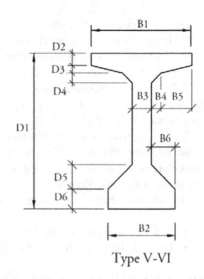

Type V-VI

FIGURE 1.12   AASHTO I-beams (PCI, 2014).

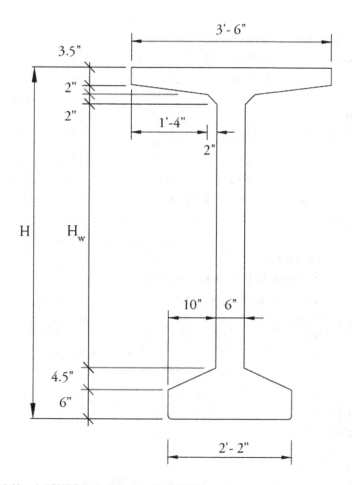

FIGURE 1.13   AASHTO bulb-T beams (PCI, 2014).

TABLE 1.4
Concrete I-beam Dimensions (inches)

Type	D1	D2	D3	D4	D5	D6	B1	B2	B3	B4	B5	B6
I	28	4	0	3	5	5	12	16	6	3	0	5
II	36	6	0	3	6	6	12	18	6	3	0	6
III	45	7	0	4.5	7.5	7	16	22	7	4.5	0	7.5
IV	54	8	0	6	9	8	20	26	8	6	0	9
V	63	5	3	4	10	8	42	28	8	4	13	10
VI	72	5	3	4	10	8	42	28	8	4	13	10

TABLE 1.5
Concrete I-beam Properties

Type	Area, in²	y_b, inches[a]	I, in⁴	w, klf
I	276	12.59	22,750	0.287
II	369	15.83	50,980	0.384
III	560	20.27	125,390	0.583
IV	789	24.73	260,730	0.822
V	1,013	31.96	521,180	1.055
VI	1,085	36.38	733,320	1.130

[a] y_b is the distance from the bottom of the beam to the centroid of the cross section.

TABLE 1.6
Concrete Bulb-T Beam Properties

Type	H, in.	H_W, in.	A, in²	I, in⁴	y_b, in.	w, klf
BT-54	54	36	659	268,077	27.63	0.686
BT-63	63	45	713	392,638	32.12	0.743
BT-72	72	54	767	545,894	36.60	0.799

TABLE 1.7
Bridge Steels

AASHTO	M270						
ASTM	A709						
Grade	36	50	50W	HPS 50W	HPS 70W	HPS 100W	HPS 100W
Shapes	All	All	All	NA	NA	NA	NA
Plates	≤ 4"	≤ 4"	≤ 4"	≤ 4"	≤ 4"	≤ 2.5"	2.5"–4"
F_y, ksi	36	50	50	50	70	100	90
F_u, ksi	58	65	70	70	85	110	100

taken from the literature (PCI, 2014). Concrete Young's modulus in AASHTO is taken to be equal to $E_c = 1,820(f'_c)^{1/2}$, where both f'_c and E_c are in units of kips per square inch (ksi).

Steel bridge girders are occasionally made from rolled steel sections, but more often are welded plate girders. Both I-girder and tub-girder sections are used, with I-girders being the most common type. Bridge steels are summarized in Table 1.7. Grade 50W is a weathering steel which performs well without paint and is often preferred due to the lower maintenance costs. HPS (high-performance steel) versions of the steels possess enhanced toughness and weldability.

TABLE 1.8
Anchor Rod Properties

Grade	F_y, ksi	F_u, ksi	Diameter, inches
36	36	58–80	½–4
55	55	75–95	½–4
105	105	125–150	½–3

Anchor rods are required in bridge structures for bearing attachment to substructures and to resist seismic loads in support diaphragms. Table 1.8 provides properties for the various grades of the preferred anchor rod material specification, ASTM F1554.

High-strength bolts used in field splices for steel I-girders, in lateral bracing member connections, and in cross-frame member connections include ASTM F3125, Grade A325 (also known as Group A) and Grade A490 (also known as Group B). Specified minimum tensile strength, F_u, is 120 ksi for Grade A325 bolts and 150 ksi for Grade A490 bolts.

Shear studs used on the top flange of steel I-girders to provide composite action with the concrete deck are required to conform to the requirements of the AASHTO/AWS D1.5M/D1.5 Bridge Welding Code and are required to have minimum specified yield and tensile strengths, $F_y = 50$ ksi and $F_u = 60$ ksi, respectively.

1.5 SOFTWARE FOR BRIDGE ENGINEERING

Software will be used frequently for the material presented in this course. Programs used will include:

- Response 2000 for concrete section-analysis, Evan Bentz, (www.ecf.utoronto.ca/~bentz/thesis.htm)
- Consec for section analysis, Robert Matthews (structware.com)
- VisualAnalysis for general structural analysis and design, Integrated Engineering Software (www.iesweb.com/edu/)
- LRFD Simon for steel bridge girder analysis and design, National Steel Bridge Alliance (www.aisc.org/nsba/design-resources/simon/)
- LEAP Bridge by Bentley (/www.bentley.com/en/products/product-line/bridge-analysis-software/openbridge-designer)

The list below summarizes other freely available resources which bridge engineers may find useful.

- FHWA Steel Bridge Design Handbook (FHWA, 2015)
- NSBA-Splice, an Excel spreadsheet for steel girder field splice design, National Steel Bridge Alliance
- NSBA Continuous Span Standards, National Steel Bridge Alliance

- PCI Bridge Design Manual (PCI, 2014)
- WSDOT BridgeLink Bridge Engineering Software, Washington State Department of Transportation

1.6 SECTION PROPERTIES

The calculation of composite section properties for bridge girders is based on theory typically learned in courses dealing with mechanics of materials, advanced steel design, and reinforced concrete, each of which is a pre-requisite to a full appreciation of this course. Deflection calculations for simple beams is covered in undergraduate structural mechanics courses. Section property calculations for composite sections are needed in all cases for modern bridge girders, both steel and prestressed concrete. Exercises for Chapter 1 focus on examples of such calculations, in addition to girder spacing and overhang issues discussed in the Introduction. For additional information on the calculation of plastic moments, the reader is referred to Appendix D6 of the LRFD BDS (AASHTO, 2020).

Recall that section property calculations for cross-sections composed of more than one material require the selection of a base material. For reinforced concrete design, concrete is selected as the base material with steel areas multiplied by the modular ratio (n) to obtain an equivalent concrete cross section. Subsequent stress calculations for the steel component require amplification by n. With steel plate girders, the reverse approach is adopted. Concrete is transformed into an equivalent steel area *via* division by the modular ratio. Subsequent stress calculations for the concrete components require division by n.

1.7 SOLVED PROBLEMS

Problem 1.1

Establish an approximate girder spacing and approximate girder depth for both the concrete and steel girder options for the Project Bridge. The deck thickness is 8.25 inches and the thickness of the haunch (the gap between the top of the girder and the bottom of the deck) is 1.75 inches.

Problem 1.2

For the concrete option on the Project Bridge, suppose a prestress force equal to 929 kips is applied to the girder as a compressive axial force at each end. Select a BT-54 girder and determine the midspan deflection due to girder self-weight and the applied axial prestress force. The prestress force is applied 4.53 inches above the bottom of the girder, thus producing not only an axial force, but also a negative moment (tension in the top of the girder) due to eccentric application. Assume the girder behaves as a simple span and is 87-ft 9-in long. Concrete strength is $f'_c = 7$ ksi. Ignore second-order effects of the axial load on deflections.

Problem 1.3

For the steel girder option of the Project Bridge, determine the composite properties (both short-term and long-term) for the elastic condition for an interior girder. Determine the plastic moment of the composite section in positive flexure (deck in compression). Use a W40×215 rolled beam girder,

8.25-inch deck, 1.75-inch haunch (distance from the top of the girder to the bottom of the deck in this case), and Grade 50 steel. Deck concrete has a specified minimum 28-day compressive strength, f'_c of 5 ksi.

PROBLEM 1.1	TEH	1/2

The following calculation shows that a girder spacing of no less than 9.057 feet is required to meet the overhang rule-of-thumb for the Project Bridge.

$$3S + 2C = 34.417 \, \text{feet}$$

$$3S + 2(0.4S) = 34.417' \rightarrow S = 9.057 \, \text{feet}$$

Select $S=9$ feet 3 inches with a corresponding $C=3$ feet 4 inches for the Project Bridge.

Define the following:

- $(D_{ss})_{PPC}$ is the superstructure depth for the prestressed girder option.
- $(D_{BM})_{PPC}$ is the girder depth for the prestressed girder option.
- $(D_{ss})_{SG}$ is the superstructure depth for the steel girder option.
- $(D_{BM})_{SG}$ is the girder depth for the steel girder option.

The superstructure depth is the girder depth plus the haunch depth plus the deck thickness.

For the precast, prestressed concrete girder option, assume continuous spans and use the information from Table 1.1.

$$\left(D_{ss}\right)_{PPC} \approx 0.040 \times 90 = 3.6 \, \text{ft} = 43 \, \text{inches}$$

Whereas $(D_{ss})_{PPC}$ is, strictly speaking, the total depth of (a) the prestressed concrete girder, (b) the haunch, and (c) the deck, experience has shown that the girder depth alone should be approximately equal to the calculated value for optimal design of prestressed girders, in the author's opinion. Without question, the estimate is just that, an estimate, and many successful projects have incorporated depths both smaller and larger.

PROBLEM 1.1	TEH	2/2

$$\left(D_{BM}\right)_{PG} \geq 43 \, \text{inches}$$

For the steel I-girder option, two criteria are specified in Table 1.1. Assess both and choose the larger value as the initial superstructure depth estimate.

For the steel I-girder plus deck criteria:

$$\left(D_{ss}\right)_{SG} \approx 0.032 \times 90 = 2.88 \, \text{ft} = 34.6 \, \text{inches}$$

$$\left(D_{BM}\right)_{SG} = 34.6 - 8.25 - 1.75 = 24.6 \, \text{inches}$$

For the steel I-girder alone:

$$\left(D_{BM}\right)_{SG} \approx 0.027 \times 90 = 2.43\,\text{ft} = 29.2\text{ inches} \leftarrow \text{controls}$$

$$\left(D_{BM}\right)_{SG} \geq 29.2\text{ inches} \leftarrow \text{controls}$$

PROBLEM 1.2	TEH	1/2

For a BT-54 girder, obtain the properties from Table 1.6.

$$A = 659\text{ in}^2$$

$$I = 268,077\text{ in}^4$$

$$y_b = 27.63\text{ inches}$$

$$w = 0.686\text{ klf}$$

The eccentricity of the applied axial force may now be computed.

$$e = 27.63 - 4.53 = 23.1\text{ inches}$$

The modulus of elasticity for concrete, E_c, is needed for deflection calculations.

$$E_C = 1,820\sqrt{7} = 4,815\text{ ksi}$$

The end moment is equal to the applied force multiplied by the eccentricity. Since this moment is negative, the deflection resulting from the moment is upward. The conjugate beam method can be used to determine the deflection due to applied end moments in a simply supported beam.

$$\Delta_{PS} = \frac{M_{PS}L^2}{8E_cI}$$

$$M_{PS} = 929 \times 23.1 = 21,460\text{ inches kips}$$

$$\Delta_{PS} = \frac{(21,460)(87.75 \times 12)^2}{8(4,815)(268,077)} = 2.30\text{ inches}\uparrow$$

PROBLEM 1.2	TEH	2/2

The deflection at midspan due to self-weight, a uniformly distributed load, may be determined from the conjugate beam method, double integration of the moment equation, or from standard textbook formulas.

$$\Delta_{SW} = \frac{5wL^4}{384E_cI} = \frac{5\left(\dfrac{0.686}{12}\right)(87.75 \times 12)^4}{384(4,815)(268,077)} = 0.71\text{ inches}\downarrow$$

The net deflection is the sum of that due to eccentricity of prestress and self-weight. The net deflection is found to be upward.

$$\Delta_{TOT} = 2.30 \text{ inches} \uparrow + 0.71 \text{ inches} \downarrow = 1.59 \text{ inches} \uparrow$$

PROBLEM 1.3	**TEH**	**1/4**

$9'3''$
$= 111''$

$49''$

$E_c = 1,820\sqrt{5}$
$= 4,070 \text{ ksi}$

$W40 \times 215, \quad A = 63.5 \text{ in}^2$
$d = 39.0 \text{ in.}$
$I_x = 16,700 \text{ in}^4$

$n = 29,000 / 4,070 = 7.13$
$3n = 21.38$

<u>Long-term Composite</u> $(3n)$

$b_{eff} = 111 / 21.38 = 5.19''$

$A = 63.5 + 5.19(8.25) = 106.3 \text{ in}^2$

$\bar{y} = \dfrac{63.5(39/2) + 42.8(44.875)}{106.3}$

$\Rightarrow \bar{y} = 29.7''$

$I = 16,700 + 63.5(29.7 - 39/2)^2$
$+ \dfrac{5.19(8.25)^3}{12} + 42.8(44.875 - 29.7)^2$

$I = 16,700 + 6,607 + 243 + 9,856$

$\Rightarrow I = 33,405 \text{ in}^4$

PROBLEM 1.3	TEH	2/4

$$S_{top} = 33,405 / (39 - 29.7) = \underline{3,592 \text{ in}^3}$$

$$S_{bott} = 33,405 / 29.7 = \underline{1,125 \text{ in}^3}$$

$$(Q/I)_{DECK} = \frac{42.8 (44.875 - 29.7)}{33,405}$$

$$= \underline{0.01944 \text{ in}^{-1}}$$

Short-term Composite (n)

$$b_{eff} = 111 / 7.13 = 15.57''$$

$$A = 63.5 + 15.57 \times 8.25 = 191.9 \text{ in}^2$$

$$\bar{y} = \frac{63.5(39/2) + 128.4(44.875)}{191.9}$$

$$\Rightarrow \bar{y} = 36.49 \text{ inches}$$

$$I = 16,700 + 63.5(36.49 - 39/2)^2$$
$$+ \frac{15.57(8.25)^3}{12} + 128.4(36.49 - 44.875)^2$$

$$I = 16,700 + 18,330 + 729 + 9,028$$

$$\Rightarrow \underline{I = 44,786 \text{ in}^4}$$

$$S_{top} = 44,786 / (39 - 36.49)$$

$$= \underline{\underline{17,843 \ in^3}}$$

$$S_{bott} = 44,786 / 36.49$$

$$= \underline{\underline{1,227 \ in^3}}$$

$$\left(Q/I\right)_{DECK} = \frac{15.57(8.25)(44.875 - 36.49)}{44,786}$$

$$= \underline{\underline{0.02405 \ in^{-1}}}$$

<u>Plastic Moment, M_p</u>

$$P_{DECK} = .85(5)(111 \times 8.25) = 3,892^k$$

$$P_{BEAM} = 50 \times 63.5 = 3,175^k$$

$$P_{BEAM} < P_{DECK} \Rightarrow \text{Plastic neutral}$$
$$\text{axis (PNA)}$$
$$\text{is in the deck.}$$

$$.85(5)(111)(a) = 3,175$$

$$\Rightarrow a = 6.73 \ inches$$

PROBLEM 1.3	TEH	4/4

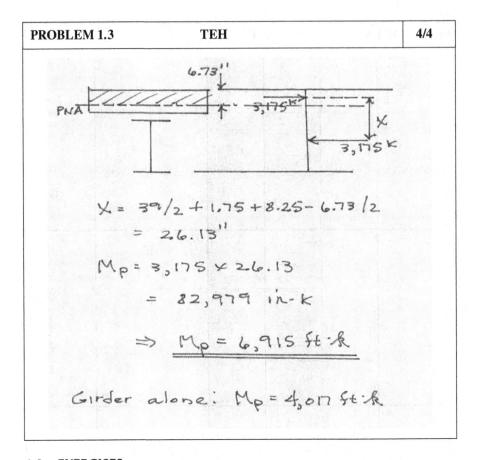

$$X = 3\%/2 + 1.75 + 8.25 - 6.73/2$$
$$= 26.13''$$

$$M_p = 3,175 \times 26.13$$
$$= 82,979 \text{ in-k}$$

$$\Rightarrow \underline{M_p = 6,915 \text{ ft-k}}$$

Girder alone: $M_p = 4,017 \text{ ft-k}$

1.8   EXERCISES

E1.1.

A three-span continuous, 129-ft 3-inches-wide interstate bridge consists of 95-ft end spans with a 156-ft center span for a total length of 346 feet. Estimate the girder depth required if (a) prestressed concrete girders are used and (b) steel I-girders are used. Also, determine the number of beams and beam spacing for each girder type.

For the prestressed girder option, use a girder spacing between 8 feet and 10 feet. For the steel I-girder option, use a girder spacing between 9 feet and 12 feet. Use a deck thickness of 8.25 inches.

E1.2.

A five-span continuous, 34-ft-wide bridge consists of 270-ft end spans and three 335-ft interior spans for a total length of 1,545 feet. Establish a preliminary cross-section (number of girders, girder spacing, and overhang dimension) configuration and steel I-girder depth for the bridge. To minimize the number of girders, use a target girder spacing of 12 feet. Deck thickness is 9 inches.

E1.3.

A 103-ft long BT-72 prestressed concrete girder consists of forty ½-inch diameter 270K-Low-Lax strands. The girder concrete strength is $f'_{ci}=6{,}000$ psi (initial strength, at release of strands). The centroid of the strand group is located 10.8 inches above the bottom of the beam. The initial pull on the strands is 75% of the tensile strength. Treat the BT-72 as a simply supported beam with self-weight, applied end compressive loads (due to the prestress force), and applied end moments (due to the eccentricity of the prestress force). This is representative of the condition for the completed beam before installation on a bridge. While the strand stress is zero at the ends and builds up to the initial pull a development length from the ends, treat the prestress force as being applied at the ends. Additionally, the strand stress experiences elastic shortening immediately and the effective prestress force would be somewhat less than $0.75f_{pu}$. For this academic problem, ignore the effects of strand development and prestress loss.

Determine the stresses (and identify whether they are tension or compression stresses) at midspan in the top and bottom of the BT-72. If the stress limits are $0.65f'_{ci}$ in compression and $0.24(f'_{ci})^{1/2}$ in tension, does the girder satisfy stress limits at midspan?

Determine the stresses (and identify whether they are tension or compression stresses) at the girder end in the top and bottom of the BT-72. If the stress limits are $0.65f'_{ci}$ in compression and $0.24(f'_{ci})^{1/2}$ in tension, does the girder satisfy stress limits at midspan?

Determine the deflection (and identify whether it is upward or downward) at midspan. Use a first-order analysis (do not consider the effect of axial compression on the deflections).

Calculate the composite moment of inertia, I_x, for positive moment (deck in compression) if the girder is to be used for a bridge with 11-foot 6-inch-girder spacing, 4,000 psi deck concrete, 8¼-inch-thick deck, and 2-inch haunch. The final girder strength is 7,000 psi. Ignore the strand area.

E1.4.

A welded steel plate girder consists of a 75-inch×½-inch web and 20-inch×1½-inch-thick flanges. All plates are made from Grade 50W steel. The girder spacing is 10 ft and the concrete deck is 8¼ inches thick. The distance from the top of the web to the bottom of the deck is 3½ inches. Deck concrete strength is 3,000 psi. The modular ratio, n, for 3,000 psi concrete is taken to be equal to 9. Determine the following properties (moment of inertia, section moduli, and shear flow) for the positive moment condition (compression in the top):

- I_x, S_{xt}, S_{xb}, Q/I_x for the girder top and bottom (girder alone)
- I_x, S_{xt}, S_{xb}, Q/I_x for the girder top and bottom (short-term composite)
- I_x, S_{xt}, S_{xb}, Q/I_x for the girder top and bottom (long-term composite)
- the web depth in compression in the elastic condition, D_c
- the web depth in compression in the plastic condition, D_{cp}
- the distance from the top of the deck to the plastic neutral axis, D_p
- the plastic moment of the composite section, M_p

E1.5.

For negative moments (tension in the top), determine the composite section properties for the girder in problem E1.4. if the area of steel in the deck is 10.00 in² with $f_y=60$ ksi, located 3½ inches from the top of the deck.

E1.6.

A welded steel plate girder consists of a 42-inch×½-inch web and 16-inch×1-inch thick flanges. All plates are made from Grade 50W steel. The girder spacing is 10 ft and the concrete deck is 8¼ inches thick. The distance from the top of the web to the bottom of the deck is 3½ inches. Deck concrete strength is 4,000 psi. The modular ratio, n, for 4,000 psi concrete is taken to be equal to 8. Determine the distance from the top of the deck to the plastic neutral axis (PNA) and the composite plastic moment in positive bending, M_p.

E1.7.

Estimate the total superstructure weight (kips per foot) for the Project Bridge. For the concrete girder option, use BT-54 girders. For the steel girder option, use W40×215 rolled steel beams. Deck thickness is 8.25 inches. Haunch thickness is 1.75 inches. Each parapet requires 0.0926 cubic yards of concrete per linear foot.

E1.8.

For the concrete girder option of the Project Bridge, estimate the total vertical reaction at the intermediate pier due to all dead loads.

2 Loads on Bridges

Bridge loads may be broadly categorized as permanent or transient, with transient loads further divided into three sub-categories: those resulting from traffic, those resulting from environmental sources, and those due to extreme events. Some of the more routinely encountered loads in each category are defined below, based on Chapter 3 of the Load and Resistance Factor Design-Bridge Design Specifications (LRFD-BDS) (AASHTO, 2020).

A. **Permanent Loads**
   - DC is the dead load of all structural components, as well as any non-structural attachments.
   - DW is the dead load of additional nonintegral wearing surfaces, future overlays, and any utilities supported by the bridge.
   - EV is the vertical earth pressure from the dead load of earth fill.
   - EH is the load due to horizontal earth pressure.
   - DD are the loads developed along the vertical sides of a deep-foundation element tending to drag it downward, typically due to consolidation of soft soils underneath embankments reducing its resistance.

B. **Transient Loads – traffic**
   - LL is the vertical gravity load due to vehicular traffic.
   - PL represents the vertical gravity load due to pedestrian traffic.
   - IM represents the dynamic load allowance to amplify LL.
   - BR is the horizontal vehicular braking force.
   - CE is the horizontal centrifugal force from vehicles on a curved roadway.

C. **Transient Loads – environment**
   - WA is the pressure due to differential water levels, stream flow, or buoyancy.
   - WS is the horizontal and vertical pressure due to wind.
   - WL is the horizontal pressure on vehicles due to wind.
   - TU is the uniform temperature change.
   - TG is the temperature gradient.
   - SE is the effect of settlement.
   - FR represents the frictional forces on sliding surfaces.

D. **Transient Loads – extreme event**
   - BL represents the intentional or unintentional forces due to blasting.
   - EQ represents loads due to earthquake ground motions.
   - CT represents horizontal impact loads due to vehicles or trains.
   - CV represents horizontal impact loads due to aberrant ships or barges.
   - IC is the horizontal static and dynamic force due to ice action.
   - SE is the effect of settlement.

DOI: 10.1201/9781003265467-2

The basic load and resistance factor design (LRFD) relationship is expressed in Equation 2.1, with the load modifier given by Equation 2.2. The load modifier, η_i, consists of three components to account for ductility, redundancy, and importance-based considerations.

$$Q_n = \sum \eta_i \gamma_i Q_i \leq \phi R_n \qquad (2.1)$$

$$\eta_i = \eta_D \eta_R \eta_I \qquad (2.2)$$

The load modifier, η_i, is to be taken to be no less than 0.95 for cases in which the use of maximum load factors, γ_i, are used. For cases in which the use of minimum load factors, γ_i, are used, η_i shall be taken no greater than 1.0.

The ductility-related modifier, η_D, is equal to 1.00 for all limit states except the Strength limit states. For the Strength limit state, η_D is equal to:

- 1.00 for conventional designs
- 1.05 for non-ductile components and connections
- 0.95 (or more) for elements designed using enhanced ductility measures beyond those required by the LRFD-BDS

The redundancy-related modifier, η_R, is equal to 1.00 for all limit states except the Strength limit states. For the Strength limit state, η_R is equal to:

- 1.00 for conventional designs
- 1.05 for non-redundant members
- 0.95 (or more) for exceptional levels of redundancy

The operational importance-related modifier, η_I, is equal to 1.00 for all limit states except the Strength limit states. For the Strength limit state, η_I is equal to:

- 1.00 for typical bridges
- 1.05 for critical or essential bridges
- 0.95 (or more) for bridges deemed to be relatively less important

The following sections provide a discussion of some of the most frequently encountered loads on bridges. See Chapter 3 for a discussion on limit states for bridges.

2.1   DEAD LOADS (DC AND DW)

DC loads for a typical bridge include the girder self-weight and associated framing (cross-frames, diaphragms, lateral bracing), the weight of the concrete deck and haunch (also known as filler; the area between the bottom of the deck and the top of the girder, filled with deck concrete but often ignored in section property calculations), sidewalks, if present, and parapets.

DW loads include an allowance for future overlay, typically about 35 psf (pounds per square foot). Also included in the DW loading is the weight of any utilities attached to the structure.

Unit weights for DC load computation are typically taken to be equal to 150 pcf (pounds per cubic foot) for concrete and 490 pcf for steel.

2.2   LIVE AND IMPACT LOADS (LL AND IM)

The traffic live load in the American Association of State Highway and Transportation Officials (AASHTO) LRFD-BDS is the HL-93 live load, consisting of a design truck with a simultaneously applied uniform design lane load. The effect of a design tandem with the simultaneously applied uniform design lane load is also to be included.

The design lane load is 0.640 klf (kips per linear foot). The design tandem consists of two 25-kip axles 5 feet apart. The design truck is a 72-kip truck with variable rear axle spacing, as depicted in Figure 2.1.

For the Fatigue limit state, a single truck is placed in a single lane. The Fatigue truck is different from the design truck only in the rear axle spacing, which is constant at 30 feet for the Fatigue truck.

For limit states other than Fatigue, multiple loaded lanes must be considered in the design of the bridge elements. To account for the rarity of maximum loading placement in multiple lanes simultaneously, a multi-presence factor, m, is applied to analytical results. This factor varies, depending on the number of lanes loaded, as shown in Table 2.1.

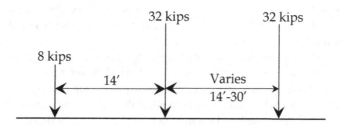

FIGURE 2.1   HL-93 design truck.

TABLE 2.1
Multi-presence factor m

Number of Lanes Loaded	Multi-presence factor, m
1	1.20
2	1.00
3	0.85
4 or more	0.65

Even though the Fatigue load, as previously mentioned, is a single truck in a single lane, the multi-presence factor, m, is not to be applied to the Fatigue limit state results. The multi-presence factor does apply for all other limit states.

Since traffic loads are not static, an impact factor, IM, is incorporated into the live load effects for the truck load only. No impact factor is included in the uniform lane loading. The LL+IM effect is the truck portion of the LL effect multiplied by (1+IM). IM is equal to:

- 0.15 for the Fatigue limit state
- 0.33 for all other limit states
- 0.75 for the design of deck joints

Bridges are designed for the number of full lanes which will physically fit on the available width. A lane is assumed to occupy a 12-foot width. Partial lanes need not be considered. For example, the Project Bridge will need to be designed for a maximum of two lanes since the available width is 32 feet (the integer portion of 32/12 is 2). Some components will be controlled by the condition for which a single lane is loaded, and other components will be controlled by the condition for which two lanes are loaded. The deck overhang, for example, will be controlled by the single-lane case, with $m = 1.2$, given that vehicles inside the exterior girder contribute no added effect to overhang deck moment. Maximum vertical load on the intermediate pier will be controlled by the two-lane case, with $m = 1.0$.

The uniform lane load and the design truck, with six feet between the wheels on a given axle, are assumed to occupy a 10-foot width and are moved laterally within the 12-foot wide lane. The design truck exterior wheel is to be placed no closer than 1.0 foot from the edge of the lane for deck design, and no closer than 2.0 feet from the edge of the lane for all other analyses.

As previously mentioned, the design live loading for bridges is one design truck plus the uniform lane load of 0.640 klf in each loaded lane. However, for girder reactions at intermediate, continuous supports, and for negative girder moments between points of contraflexure, the AASHTO LRFD-BDS requires that 90% of the effect of two trucks plus 90% of the effect of the lane load in each loaded lane be considered as well.

To summarize the application of the HL-93 live load on bridges, the extreme effect is the larger of the following:

a) the effect of the design tandem combined with the effect of the design lane load, or
b) the effect of one design truck with the variable axle spacing combined with the effect of the design lane load, or
c) for negative moment between points of contraflexure and for reactions at interior piers only, 90 percent of the effect of two design trucks combined with 90 percent of the effect of the design lane load.

For load placement in (c), the trucks are to be placed with a minimum of 50.0 ft between the lead axle of the second truck and the rear axle of the first truck, and the distance between the rear 32.0-kip axles of each truck is to be 14.0 ft. Since case (c) applies only for negative moment and pier reactions, the two design trucks are to be placed in adjacent spans. It would thus seem that it is seldom, if ever, necessary to place two trucks in the same span for the HL-93 live loading.

2.3 BRAKING FORCES (BR)

Braking forces occur when traffic slows on a bridge deck. The design braking force is taken to be applied 6 feet above the surface of the deck, is placed in all lanes carrying traffic in the same direction, and is equal to the larger of:

a) 25 percent of the axle weights of the design truck or design tandem, or
b) 5 percent of the design truck plus lane load, or
c) 5 percent of the design tandem plus lane load

Braking forces should be distributed to substructures (abutments and piers) according to the relative stiffness of each substructure.

With expansion joints at each abutment, a typical assumption would be that the piers (bents) carry 100% of the braking forces. With integral abutments, the abutments are likely to carry a large proportion of the braking forces. In such cases, a lower bound abutment stiffness might be useful in estimating the fraction of braking forces to be assigned to the piers, with an upper bound abutment stiffness used to determine braking forces assigned to the abutments.

2.4 CENTRIFUGAL FORCES (CE)

Centrifugal forces occur when traffic moves on a bridge which is curved in plan. With effects from opposing lanes on a two-directional, two-lane bridge offsetting one another, one lane of centrifugal force (with $m = 1.2$) may suffice. However, it may also be necessary to consider the possibility of a bridge becoming one-directional in the future and designing for such a condition with multiple lanes of centrifugal force, and an appropriate multi-presence factor.

The centrifugal force (F_{CE}) imparted to a bridge is a horizontal force to be applied transversely to the direction of travel (away from the center of curve), six feet above the deck, and is given by Equation 2.3.

$$F_{CE} = CW \tag{2.3}$$

$$C = f\,\frac{v^2}{gR} \tag{2.4}$$

where W is the weight of the truck (72 kips) or tandem (50 kips). The design speed, v, must be expressed in units of ft/s, g is the acceleration of gravity (32.2 ft/s^2), and R is the radius of curvature of the loaded lane (ft). The factor, f, is 1.0 for the Fatigue limit state and 4/3 for all other limit states.

Centrifugal forces alter the weight distribution on the design truck wheels. For a stationary vehicle, 50% of the weight goes to each wheel line. It may be shown that the distribution of vehicle weight to the wheel lines is given by Equations 2.5 through 2.10, with reference to Figure 2.2 for the definition of the various force components. The effect can be significant and, although often ignored in the computation of live load distribution factors, may need to be considered for curved structures.

$$F_{N2}\big/_W = C\left[\left[\left(\frac{h}{b}\right)\cos\phi + \left(\frac{1}{2}\right)\sin\phi\right] + \left[\left(\frac{1}{2}\right)\cos\phi - \left(\frac{h}{b}\right)\sin\phi\right]\right] \qquad (2.5)$$

$$F_{N1}\big/_W = \cos\phi + C\left(\sin\phi\right) - \left(F_{N2}\big/_W\right) \qquad (2.6)$$

$$F_{T1}\big/_W = \frac{C\left(\cos\phi\right) - \sin\phi}{\left(1 + F_{N2}\big/_{F_{N1}}\right)} \qquad (2.7)$$

$$F_{T2}\big/_W = \left(F_{T1}\big/_W\right)\cdot\left(F_{N2}\big/_{F_{N1}}\right) \qquad (2.8)$$

$$F_{V1}\big/_W = \left(F_{N1}\big/_W\right)\cos\phi - \left(F_{T1}\big/_W\right)\left(\sin\phi\right) \qquad (2.9)$$

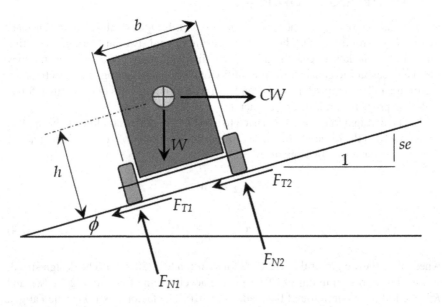

FIGURE 2.2   Centrifugal forces.

$$F_{V1}\Big/_W = \left(F_{N1}\Big/_W\right)\cos\phi - \left(F_{T1}\Big/_W\right)(\sin\phi) \qquad (2.10)$$

F_{V1}/W and F_{V2}/W are the relative vertical components of load distributed to each wheel line and may be useful in the lever rule or in the rigid cross-section methods for live load distribution factor calculations (see Chapter 5). Note that, for the design truck, $h=b=6$ feet.

Suppose $R=730$ ft and the superelevation, $SE=0.08$, with the design speed equal to 50 mph. Solving the equations, it may be advisable in such a case to use a wheel load distribution of 0.28 (inner) and 0.72 (outer), rather than the usual 0.50 (inner) and 0.50 (outer), for live load distribution.

Note as well that the typical coefficient of friction, μ, between dry pavement and average tires is 0.80, between wet pavement and a grooved tire is 0.70, and between wet pavement and a smooth tire is 0.40. In all three cases, the friction is sufficient to resist the tangential forces since $F_{T1}/F_{N1} = F_{T2}/F_{N2} = 0.220$. This means that a vehicle travelling at the 50-mph design speed, with the specified $R=730$ ft and $SE=0.08$, is unlikely to slide.

2.5 WIND LOADS (WS AND WL)

Wind loads in the AASHTO LRFD-BDS are based on 3-second gust wind speeds. This was not always the case, as previous versions used fastest-mile wind speed as the design basis. It is important not to mix codes and specifications in structural design for this, and other, reasons. Load factors and resistance factors vary among design standards.

For the Strength III limit state, wind speeds are taken from maps in the AASHTO LRFD-BDS. For the Service IV limit state, the wind speed is taken as 75% of that for the Strength III limit state. Other limit states for which wind loading is applied include Service I, for which a wind speed of 70 mph is used, and Strength V, for which the design wind speed is 80 mph. See Chapter 3 for a detailed discussion of limit states.

The basic horizontal wind pressure equation is given here in Equation 2.11. The velocity, V, must be expressed in mph (miles per hour). The resulting pressure is in ksf (kips per square foot). The gust factor, G, is 1.0 for structures other than sound barriers. The drag coefficient, C_D, is 1.3 for I-girder superstructures and 1.6 for bridge substructures. The surface exposure and elevation coefficient, K_Z, accounts for height above ground (or water) and surface roughness conditions. Maps in the AASHTO LRFD-BDS or Equations 2.12 through 2.14 may be used to determine K_Z for Strength III and Service IV limit states. For all other limit states, K_Z is equal to 1.0. The height above ground or water, Z, is never to be taken less than 33 feet. Wind loads on substructures are to be computed using the superstructure height above ground or water, unless otherwise approved by the Owner.

$$P_Z = 2.56 \times 10^{-6} \times V^2 K_Z G C_D \qquad (2.11)$$

$$K_z(B) = \frac{\left[2.5\ln\left(\dfrac{Z}{0.9834}\right) + 6.87\right]^2}{345.6} \tag{2.12}$$

$$K_z(C) = \frac{\left[2.5\ln\left(\dfrac{Z}{0.0984}\right) + 7.35\right]^2}{478.4} \tag{2.13}$$

$$K_z(D) = \frac{\left[2.5\ln\left(\dfrac{Z}{0.0164}\right) + 7.65\right]^2}{616.1} \tag{2.14}$$

Ground surface roughness 'B' generally refers to "terrain with numerous closely spaced obstructions having the size of single-family dwellings or larger" (AASHTO, 2020).

Ground surface roughness 'C' generally refers to "open terrain with scattered obstructions having heights generally less than 33.0 ft, including flat open country and grasslands" (AASHTO, 2020).

Ground surface roughness 'D' generally refers to "flat, unobstructed areas and water surfaces", including "smooth mud flats, salt flats, and unbroken ice" (AASHTO, 2020).

The reader is referred to Section 3.8 of the LRFD-BDS for detailed discussions on the determination of appropriate ground surface roughness designations.

Multiple angles of wind direction are typically investigated. For the case in which the wind is perpendicular to the bridge (and only for this case), a vertical (upward) pressure of 20 psf (Strength III) or 10 psf (Service IV) is applied to the bridge deck at one-quarter of the distance from the windward edge of the bridge. This load acts concurrently with the horizontal wind pressures and is not applied for limit states other than Strength III and Service IV.

In addition to wind load on the structure, WS, AASHTO also requires consideration of wind load on the live load, WL. The wind load on traffic is taken to be 0.10 klf transverse to traffic and 0.04 klf in the direction of traffic, both applied 6 feet above the deck surface.

2.6 COLLISION LOADS (CT AND CV)

For substructures located closer than 30 feet (typical clear zone) to the roadway edge, and with no independent barrier designed to withstand vehicle collision, the substructure components must be designed to withstand such collision loads. Vehicle collision loading (CT) is defined as a 600-kip load, five feet above ground at an angle of between 0 and 15 degrees with the pavement edge. This vehicle collision load is based on crash tests of rigid columns subjected to an 80-kip vehicle at a speed of 50 mph.

As an alternative to designing substructures within 30 feet of the roadway for the vehicular collision load, an independent barrier protecting the substructure may be used. Such barriers may be any of the following:

- An embankment;
- A structurally independent, crashworthy, ground-mounted 54-inch-high barrier, located within 10 ft of the component being protected;
- A 42-inch high barrier located at more than 10 ft from the component being protected.

Such barriers are required to be structurally and geometrically capable of surviving the crash test for Test Level 5 as specified in Section 13 of the AASHTO LRFD-BDS.

Bridge parapets and decks are designed for vehicular crash load as well. Table 2.2 summarizes the various test levels specified in Section 13 of the AASHTO-LRFD-BDS.

The parameters are defined as follows:

- F_t=transverse impact force applied at height H_e, kips
- F_l=longitudinal impact force applied at height H_e, kips
- F_v=vertical impact force applied at height H_e, kips
- H=height of wall, ft
- L_t=longitudinal length of distribution of horizontal impact force, ft
- L_v=longitudinal length of distribution of vertical impact force, ft

Bridge components in navigable waterways must be designed for vessel collision (CV) loading or protected by barriers such as dolphins, fenders, or berms. The interested reader is referred to Section 3.14 of the AASHTO-LRFD-BDS for guidance on vessel collision load determination.

TABLE 2.2
Crash Load Parameters for Various Test Levels

Parameter	TL-1	TL-2	TL-3	TL-4	TL-5	TL-6
F_t, transverse (kips)	13.5	27.0	54.0	54.0	124.0	175.0
F_L, longitudinal (kips)	4.5	9.0	18.0	18.0	41.0	58.0
F_v, vertical (kips)	4.5	4.5	4.5	18.0	80.0	80.0
L_t and L_L (feet)	4.0	4.0	4.0	3.5	8.0	8.0
L_v (feet)	18.0	18.0	18.0	18.0	40.0	40.0
H_e (min, inches)	18.0	20.0	24.0	32.0	42.0	56.0
Min rail height, H (ft)	27.0	27.0	27.0	32.0	42.0	90.0

2.7 TEMPERATURE LOADS (TU)

For thermal expansion and contraction requirements, sites are classified as either 'moderate climate' or 'cold climate' conditions in the AASHTO-BDS. The reader is referred to Section 3.12 in the AASHTO-BDS for maps.

Figure 2.3 depicts a scenario for a five-span bridge with piers of varying height. Although some engineers assume the 'center of stiffness' to be midway between the bridge ends, such an assumption can result in serious error. For each substructure element contributing to the longitudinal stiffness of the system, Equation 2.15 may be used to establish the location of the center of stiffness – the stationary point on the superstructure from which expansion and contraction due to temperature variation take place. Although these estimates are approximate due to the difficulty in estimating substructure stiffness (K_i) values (particularly for abutments without expansion joints; for abutments with expansion joints, the abutment stiffness may be taken to be equal to zero), they are often preferable to traditional assumptions. The center of stiffness is located L_{A1} from abutment number 1. The distance from abutment 1 to each substructure is x_i. Equation 2.16 gives the total thermal movement required at a given substructure location. As shown in Figure 2.3, L_i is the distance from the center of stiffness to the substructure in question.

$$L_{A1} = \frac{\sum K_i \cdot x_i}{\sum K_i} \tag{2.15}$$

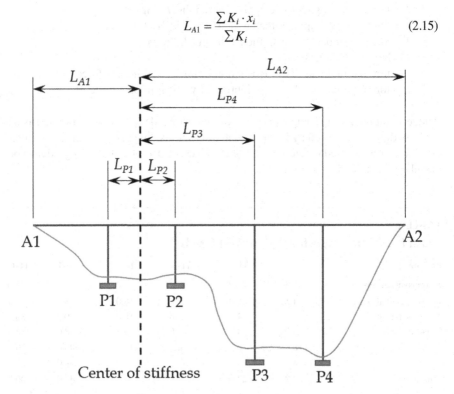

FIGURE 2.3   Thermal expansion calculation schematic.

$$\left(\Delta_{TU}\right)_i = \alpha\left(\delta T\right)L_i \qquad (2.16)$$

For bridges located in moderate climates, the design temperature range, δT is as follows:

- $\delta T = 0°–120°$ for steel girder superstructures
- $\delta T = 10°–80°$ for concrete girder superstructures

For bridges located in cold climates, the design temperature range, δT is as follows:

- $\delta T = -30°–+120°$ for steel girder superstructures
- $\delta T = 0°–80°$ for concrete girder superstructures

The coefficient of thermal expansion, α, is 0.0000060/°F for concrete girder super-structures and 0.0000065/°F for steel girder superstructures.

Equation 2.16 gives the total theoretical movement, which may then be broken into expansion and contraction components. In an idealized situation, one-half of the total movement would be expansion and one-half would be contraction. To account for variations in temperature at the time of structure completion and inaccuracies inherent in such computations, the AASHTO LRFD-BDS adopts a load factor of 1.2 on TU deformation loads at both the Service and Strength limit states (see Chapter 3, Table 3.1 of this book). Therefore, the design total movement may be taken as 1.2 times that given by Equation 2.16, resulting in expansion and contraction require-ments each equal to 60% of that given by Equation 2.16. Engineers may wish to add an additional contingency to calculated movements, given the uncertainties involved in the estimates. The method has been used successfully on many projects. Note, however, that for the design of reinforced elastomeric bearings, Section 14.7.5.3.2 of the LRFD-BDS requires that 65% (rather than 60%) of the total movement be used in the design of such bearings.

2.8 EARTHQUAKE LOADS (EQ)

Seismic loading in the AASHTO LRFD-BDS is defined as a geometric-mean-based, uniform hazard, response spectrum. This is in contrast to building design in accordance with ASCE 7-16, which defines earthquake ground motion for design as the maximum-direction-based, risk-targeted, response spectrum. Such definitions in design codes and specifications change frequently, and engineers would be well served by keeping abreast of changes in the nature of seismic load definitions in AASHTO, in ASCE 7, and in any applicable design standard.

A Poisson probability distribution is typically used as the model defining relation-ships among exposure time, t, probability of exceedance, PE, and mean recurrence interval, MRI. The model is given here in Equation 2.17.

$$MRI = \frac{-t}{\ln\left(1 - PE\right)} \qquad (2.17)$$

The shape of the design response spectrum is depicted in Figure 2.4. Determination of the control points (A_S, S_{DS}, S_{D1}, T_S, T_O) on the design response spectrum requires the determination of an appropriate site class for the project. A response spectrum is a plot of pseudo-spectral acceleration, *PSA*, *versus* structure period, *T*.

The code-based shape is entirely determined by the control points. However, current trends in building design may result in a 22-point design response spectrum with site effects incorporated automatically. This provided further impetus for engineers to be fully aware of the nature and basis of design response spectra in the governing, current design specification.

As of December 2020, the design ground motion in AASHTO is that having a 7% probability of exceedance in 75 years. With $t = 75$ years and $PE = 0.07$, Equation 2.17 gives $MRI = 1,033$ years. The USGS Unified Hazard Tool will be useful in determining uniform hazard, geometric-mean-based *PSA* values for a site with a given latitude and longitude. The application URL is listed below.

https://earthquake.usgs.gov/hazards/interactive/

Site class determination is based on either average shear wave velocity in the upper 100 feet of the subsurface profile, or, more frequently, on the average standard penetration test blow count in the upper 100 feet. The shear wave velocity averaged over the upper 100 feet (30 meters) of the soil profile is termed V_{S30}. It is important

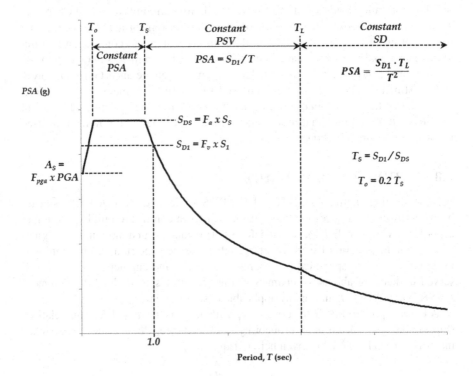

FIGURE 2.4   AASHTO design response spectrum control points.

to recognize that a velocity-based average is not computed in the same way as a distance-based average is calculated. The correct calculation of V_{S30} is given in Equations 2.18 and 2.19. Only the top 30 meters of the profile is used to calculate V_{S30}. Two calculations are necessary: (1) consider all layers in the top 30 meters and (2) consider only the cohesionless layers in the top 30 meters.

$$\left(V_{S30}\right)_1 = \frac{\sum d_i}{\sum \dfrac{d_i}{V_{Si}}}, \text{all layers} \tag{2.18}$$

$$\left(V_{S30}\right)_2 = \frac{\sum d_i}{\sum \dfrac{d_i}{V_{Si}}}, \text{include cohesionless layers ONLY} \tag{2.19}$$

Whereas a true average blow count would be computed differently, the same format is to be used when blow count data is selected as the basis for site classification, with V_{Si} replaced by N_i (blow count) in Equations 2.18 and 2.19. N_i is never to be taken larger than 100 blows per foot and should generally be taken as equal to 100 blows per foot for rock, when such occurs at a depth of less than 100 feet.

Table 2.3 summarizes criteria used to establish the appropriate site class for a project. In Table 2.3, S_u is the undrained shear strength.

The long transition period, T_L, is not available in the USGS Unified Hazard Tool but is available in the ATC Hazard by Location online Tool. The ATC application produces design *PSA* control points for buildings and should not be used to determine parameters other than T_L for bridges at this time. The tool URL is shown below.

https://hazards.atcouncil.org/

Site factors are to be applied to *PSA* values at the B/C boundary. Note that the code-based site factors are typically 1.00 for Site Class B in current codes and specifications for bridges. *PGA*, S_S, and S_1 are mapped accelerations at the B/C boundary. As Figure 2.4 shows, these three pseudo-spectral accelerations are multiplied by site factors, F_{PGA}, F_a, and F_v, respectively, to define the design response spectrum.

TABLE 2.3
Site Class Definitions

Site Class	V_{S30}, ft/sec	N, blows/ft	S_u, psf
A. Hard Rock	>5,000	NA	NA
B. Rock	2,500–5,000	NA	NA
C. Very Dense Soil/Soft Rock	1,200–2,500	>50	>2,000
D. Stiff Soil	600–1,200	15–50	1,000–2,000
E. Soft Clay Soil	<600	<15	<1,000
F. Soils Requiring Site Response	Liquefaction, peats, highly sensitive or plastic clays		

Code-based site factors from the AASHTO LRFD-BDS are summarized in Tables 2.4 through 2.6. However, these site factors may not be suitable for areas with deep soil profiles, such as are found in the Mississippi Embayment (ME) of the New Madrid seismic zone (NMSZ), among other regions. Research performed in the past 15 years or so has produced site factors for deep soil sites much different from code-based values. Tables 2.7 and 2.8 summarize such findings from one such study (Hashash et al., 2008). Tables 2.9 through 2.12 summarize results from another (Malekmohammadi and Pezeshk, 2014). Both studies distinguish between so-called "uplands" and "lowlands" sites within the ME.

Engineers involved in structural design on deep soil sites may wish to consider alternative site factors, such as those presented here.

The multitude of seismic design spectrum options is evident through browsing the USGS Seismic Design Web Services page:

https://earthquake.usgs.gov/ws/designmaps/

TABLE 2.4
AASHTO F_{pga} Site Coefficient

	F_{pga}				
	PGA Range of Applicability				
Site Class	0.10	0.20	0.30	0.40	0.50
A	0.80	0.80	0.80	0.80	0.80
B	1.00	1.00	1.00	1.00	1.00
C	1.20	1.20	1.10	1.00	1.00
D	1.60	1.40	1.20	1.10	1.00
E	2.50	1.70	1.20	0.90	0.90

TABLE 2.5
AASHTO F_a Site Coefficient

	F_a				
	S_S Range of Applicability				
Site Class	0.25	0.50	0.75	1.00	1.25
A	0.80	0.80	0.80	0.80	0.80
B	1.00	1.00	1.00	1.00	1.00
C	1.20	1.20	1.10	1.00	1.00
D	1.60	1.40	1.20	1.10	1.00
E	2.50	1.70	1.20	0.90	0.90

TABLE 2.6
AASHTO F_v Site Coefficient

	F_v				
	S_1 Range of Applicability				
Site Class	0.10	0.20	0.30	0.40	0.50
A	0.80	0.80	0.80	0.80	0.80
B	1.00	1.00	1.00	1.00	1.00
C	1.70	1.60	1.50	1.40	1.30
D	2.40	2.00	1.80	1.60	1.50
E	3.50	3.20	2.80	2.40	2.40

TABLE 2.7
Site Factor F_a from Hashash et al. (2008)

	$S_S=0.25$		$S_S=0.50$		$S_S=0.75$		$S_S=1.00$		$S_S\geq1.25$	
Thickness (meters)	Up	Low	Up	Low	Up	Low	Up	Low	Up	Low
30	1.46	1.41	1.32	1.27	1.18	1.13	1.11	1.06	1.06	1.01
100	1.41	1.31	1.27	1.17	1.13	1.03	1.06	0.96	1.01	0.91
200	1.36	1.21	1.22	1.07	1.08	0.93	1.01	0.86	0.96	0.81
300	1.31	1.11	1.17	0.97	1.03	0.83	0.96	0.76	0.91	0.71
500	1.27	1.06	1.13	0.92	0.99	0.78	0.92	0.71	0.87	0.66
1000	1.23	1.04	1.09	0.90	0.95	0.76	0.88	0.70	0.83	0.64

'Up' = Uplands; 'Low' = Lowlands.

TABLE 2.8
Site Factor F_v from Hashash et al. (2008)

	$S_1=0.10$		$S_1=0.20$		$S_1=0.30$		$S_1=0.40$		$S_1\geq0.50$	
Thickness (meters)	Up	Low	Up	Low	Up	Low	Up	Low	Up	Low
30	2.40	2.40	2.00	2.00	1.80	1.80	1.60	1.60	1.50	1.50
100	2.70	2.55	2.30	2.15	2.10	1.95	1.95	1.75	1.80	1.65
200	2.85	2.67	2.45	2.27	2.25	2.07	2.08	1.87	1.91	1.77
300	2.95	2.77	2.55	2.37	2.37	2.17	2.18	1.97	2.01	1.87
500	3.00	2.82	2.60	2.42	2.42	2.22	2.23	2.02	2.06	1.92
1000	3.05	2.87	2.65	2.47	2.47	2.27	2.28	2.07	2.08	1.97

'Up' = Uplands; 'Low' = Lowlands.

TABLE 2.9

Uplands Site Factor F_a from Malekmohammadi and Pezeshk (2014)

V_{S30}, m/s	Site Class	Depth, m	$S_S \leq 0.25$	$S_S = 0.50$	$S_S = 0.75$	$S_S = 1.00$	$S_S \geq 1.25$
560	C	30	1.509	1.228	1.049	0.923	0.829
		70	1.624	1.285	1.081	0.940	0.837
		140	1.618	1.250	1.036	0.892	0.788
		400	1.362	1.011	0.819	0.693	0.604
		750	1.069	0.774	0.617	0.517	0.447
270	D	30	1.528	1.057	0.803	0.647	0.543
		70	1.660	1.117	0.836	0.666	0.553
		140	1.667	1.095	0.807	0.637	0.525
		400	1.421	0.896	0.646	0.501	0.408
		750	1.123	0.691	0.491	0.377	0.304
180	E	30	1.451	0.900	0.638	0.490	0.396
		70	1.581	0.954	0.666	0.505	0.405
		140	1.592	0.938	0.645	0.485	0.385
		400	1.362	0.771	0.518	0.383	0.301
		750	1.079	0.595	0.394	0.288	0.225

TABLE 2.10

Uplands Site Factor F_v from Malekmohammadi and Pezeshk (2014)

V_{S30}, m/s	Site Class	Depth, m	$S_1 \leq 0.10$	$S_1 = 0.20$	$S_1 = 0.30$	$S_1 = 0.40$	$S_1 \geq 0.50$
560	C	30	3.304	2.841	2.550	2.340	2.179
		70	4.428	3.862	3.496	3.227	3.017
		140	5.630	4.947	4.498	4.165	3.904
		400	4.171	3.708	3.394	3.158	2.971
		750	3.559	3.181	2.921	2.724	2.567
270	D	30	3.771	2.753	2.176	1.803	1.543
		70	4.383	3.245	2.586	2.155	1.853
		140	4.974	3.711	2.970	2.483	2.140
		400	4.397	3.318	2.674	2.246	1.943
		750	4.170	3.164	2.558	2.153	1.866
180	E	30	3.604	2.341	1.703	1.327	1.083
		70	4.007	2.640	1.937	1.518	1.244
		140	4.390	2.914	2.148	1.688	1.387
		400	4.099	2.753	2.042	1.613	1.330
		750	4.017	2.713	2.019	1.598	1.320

TABLE 2.11
Lowlands Site Factor F_a from Malekmohammadi and Pezeshk (2014)

V_{S30}, m/s	Site Class	Depth, m	$S_S \leq 0.25$	$S_S = 0.50$	$S_S = 0.75$	$S_S = 1.00$	$S_S \geq 1.25$
560	C	30	2.156	1.844	1.622	1.457	1.330
		70	2.159	1.780	1.532	1.355	1.222
		140	2.068	1.653	1.398	1.221	1.091
		400	1.668	1.273	1.048	0.898	0.790
		750	1.283	0.953	0.772	0.654	0.570
270	D	30	2.182	1.586	1.242	1.022	0.870
		70	2.207	1.546	1.185	0.960	0.808
		140	2.131	1.448	1.090	0.872	0.727
		400	1.740	1.129	0.827	0.649	0.533
		750	1.348	0.851	0.614	0.476	0.388
180	E	30	2.072	1.351	0.987	0.773	0.636
		70	2.103	1.321	0.944	0.729	0.592
		140	2.035	1.241	0.871	0.663	0.534
		400	1.668	0.971	0.663	0.496	0.393
		750	1.296	0.743	0.493	0.365	0.286

TABLE 2.12
Lowlands Site Factor F_v from Malekmohammadi and Pezeshk (2014)

V_{S30}, m/s	Site Class	Depth, m	$S_1 \leq 0.10$	$S_1 = 0.20$	$S_1 = 0.30$	$S_1 = 0.40$	$S_1 \geq 0.50$
560	C	30	3.366	3.131	2.944	2.792	2.666
		70	4.494	4.175	3.921	3.716	3.546
		140	5.702	5.290	4.965	4.702	4.486
		400	4.215	3.918	3.682	3.490	3.332
		750	3.593	3.344	3.144	2.982	2.848
270	D	30	3.842	3.035	2.512	2.150	1.888
		70	4.449	3.509	2.901	2.482	2.178
		140	5.038	3.968	3.278	2.803	2.459
		400	4.443	3.507	2.901	2.482	2.179
		750	4.210	3.326	2.753	2.358	2.070
180	E	30	3.671	2.581	1.966	1.583	1.325
		70	4.067	2.854	2.173	1.748	1.462
		140	4.446	3.116	2.370	1.906	1.594
		400	4.142	2.909	2.216	1.783	1.492
		750	4.055	2.852	2.173	1.749	1.464

It is important for engineers to know the basis (required *MRI*) and nature (risk-targeted or uniform hazard, geomean or maximum direction, three-point or twenty-two-point, etc.) of the required design response spectra, and to use the appropriate tools (site factors if required, web-based applications, etc.) to generate such spectra.

A third, more recent, alternative for site factors in the New Madrid seismic zone has been summarized in the Korean Society of Civil Engineers Journal (Moon et al., 2016). Tables 2.13 through 2.18 summarize the recommended site factors. The developed site factors are compared to National Earthquake Hazards Reduction Program (NEHRP) site factors in the report and in the Tables.

Given the multitude of uncertainties in site response analysis and in geotechnical investigations to determine soil properties, it may be more prudent to select the

TABLE 2.13

Mississippi Embayment Site Factor F_a from Moon et al. (2016) –Site Class C

H, m	S_S	NEHRP	Upland	Lowland
30	0.25	1.20	1.80	1.72
30	0.50	1.20	1.75	1.68
30	0.75	1.10	1.66	1.62
30	1.00	1.00	1.57	1.49
30	1.25	1.00	1.54	1.47
100	0.25	1.20	1.73	1.69
100	0.50	1.20	1.68	1.65
100	0.75	1.10	1.59	1.57
100	1.00	1.00	1.51	1.46
100	1.25	1.00	1.48	1.43
200	0.25	1.20	1.67	1.63
200	0.50	1.20	1.61	1.59
200	0.75	1.10	1.53	1.51
200	1.00	1.00	1.46	1.40
200	1.25	1.00	1.42	1.37
300	0.25	1.20	1.62	1.58
300	0.50	1.20	1.56	1.53
300	0.75	1.10	1.48	1.45
300	1.00	1.00	1.40	1.34
300	1.25	1.00	1.37	1.31
500	0.25	1.20	1.54	1.52
500	0.50	1.20	1.49	1.46
500	0.75	1.10	1.40	1.35
500	1.00	1.00	1.31	1.24
500	1.25	1.00	1.28	1.21
1,000	0.25	1.20	1.39	1.38
1,000	0.50	1.20	1.32	1.29
1,000	0.75	1.10	1.24	1.17
1,000	1.00	1.00	1.14	1.07
1,000	1.25	1.00	1.11	1.05

TABLE 2.14
Mississippi Embayment Site Factor F_a from Moon et al. (2016) – Site Class D

H, m	S_S	NEHRP	Upland	Lowland
30	0.25	1.60	1.65	1.45
30	0.50	1.40	1.48	1.34
30	0.75	1.20	1.21	1.07
30	1.00	1.10	1.06	0.94
30	1.25	1.00	0.97	0.84
100	0.25	1.60	1.54	1.35
100	0.50	1.40	1.42	1.24
100	0.75	1.20	1.14	0.97
100	1.00	1.10	1.00	0.84
100	1.25	1.00	0.91	0.75
200	0.25	1.60	1.49	1.28
200	0.50	1.40	1.39	1.19
200	0.75	1.20	1.12	0.92
200	1.00	1.10	0.98	0.79
200	1.25	1.00	0.88	0.69
300	0.25	1.60	1.48	1.25
300	0.50	1.40	1.37	1.14
300	0.75	1.20	1.09	0.87
300	1.00	1.10	0.96	0.75
300	1.25	1.00	0.86	0.66
500	0.25	1.60	1.46	1.20
500	0.50	1.40	1.33	1.09
500	0.75	1.20	1.05	0.82
500	1.00	1.10	0.92	0.70
500	1.25	1.00	0.83	0.64
1,000	0.25	1.60	1.41	1.15
1,000	0.50	1.40	1.27	1.04
1,000	0.75	1.20	0.97	0.77
1,000	1.00	1.10	0.84	0.66
1,000	1.25	1.00	0.80	0.62

worst-case site factors from tabulated values, rather than interpolating. For example, suppose the Moon et al. (2016) site factors are being used at a lowland, Class E, Mississippi Embayment site with $S_1 = 0.45$ and a profile depth, $H = 450$ meters. Applicable tabulated values bounding the problem are:

- $S_1 = 0.40$, $H = 300$ meters $\longrightarrow F_v = 2.68$
- $S_1 = 0.50$, $H = 300$ meters $\longrightarrow F_v = 2.33$
- $S_1 = 0.40$, $H = 500$ meters $\longrightarrow F_v = 2.83$
- $S_1 = 0.50$, $H = 500$ meters $\longrightarrow F_v = 2.54$

TABLE 2.15

Mississippi Embayment Site Factor F_a from Moon et al. (2016) – Site Class E

H, m	S_S	NEHRP	Upland	Lowland
30	0.25	2.50	1.62	1.46
30	0.50	1.70	1.10	0.99
30	0.75	1.20	0.92	0.87
30	1.00	0.90	0.79	0.74
30	1.25	0.90	0.69	0.64
100	0.25	2.50	1.56	1.44
100	0.50	1.70	1.05	0.95
100	0.75	1.20	0.87	0.81
100	1.00	0.90	0.75	0.69
100	1.25	0.90	0.66	0.58
200	0.25	2.50	1.52	1.44
200	0.50	1.70	1.03	0.95
200	0.75	1.20	0.87	0.81
200	1.00	0.90	0.75	0.69
200	1.25	0.90	0.66	0.58
300	0.25	2.50	1.48	1.42
300	0.50	1.70	1.03	0.95
300	0.75	1.20	0.86	0.81
300	1.00	0.90	0.75	0.69
300	1.25	0.90	0.66	0.58
500	0.25	2.50	1.44	1.40
500	0.50	1.70	1.00	0.95
500	0.75	1.20	0.83	0.81
500	1.00	0.90	0.72	0.68
500	1.25	0.90	0.64	0.58
1,000	0.25	2.50	1.38	1.34
1,000	0.50	1.70	0.94	0.91
1,000	0.75	1.20	0.79	0.76
1,000	1.00	0.90	0.69	0.63
1,000	1.25	0.90	0.62	0.58

Two-way interpolation would yield $F_v = 2.64$. While this value would likely be permissible, a more logical solution might be to take $F_v = 2.83$, the largest of the four tabulated values surrounding the prescribed site conditions.

Based on the *PSA* at a 1-second period, S_{D1}, a "Seismic Zone" is assigned in the LRFD-BDS for force-based seismic design.

Based on the *PSA* at a 1-second period, S_{D1}, a "Seismic Design Category" is assigned in the LRFD-GS for displacement-based seismic design.

- $S_{D1} \leq 0.15$ $\longrightarrow$ Seismic Zone 1 $\longrightarrow$ Seismic Design Category A
- $0.15 < S_{D1} \leq 0.30$ $\longrightarrow$ Seismic Zone 2 $\longrightarrow$ Seismic Design Category B

TABLE 2.16
Mississippi Embayment Site Factor F_v from Moonet al. (2016) – Site Class C

H, m	S_1	NEHRP	Upland	Lowland
30	0.10	1.70	1.42	1.40
30	0.20	1.60	1.40	1.39
30	0.30	1.50	1.36	1.38
30	0.40	1.40	1.35	1.29
30	0.50	1.30	1.34	1.24
100	0.10	1.70	1.59	1.76
100	0.20	1.60	1.53	1.73
100	0.30	1.50	1.48	1.71
100	0.40	1.40	1.44	1.55
100	0.50	1.30	1.41	1.51
200	0.10	1.70	1.82	2.00
200	0.20	1.60	1.72	1.91
200	0.30	1.50	1.61	1.80
200	0.40	1.40	1.55	1.66
200	0.50	1.30	1.54	1.65
300	0.10	1.70	1.98	2.22
300	0.20	1.60	1.83	2.03
300	0.30	1.50	1.67	1.81
300	0.40	1.40	1.55	1.66
300	0.50	1.30	1.54	1.65
500	0.10	1.70	1.98	2.22
500	0.20	1.60	1.83	2.03
500	0.30	1.50	1.67	1.81
500	0.40	1.40	1.55	1.66
500	0.50	1.30	1.54	1.65
1,000	0.10	1.70	1.98	2.22
1,000	0.20	1.60	1.83	2.03
1,000	0.30	1.50	1.67	1.81
1,000	0.40	1.40	1.55	1.66
1,000	0.50	1.30	1.54	1.65

- $0.30 < S_{DI} \leq 0.50$ $\longrightarrow$ Seismic Zone 3 $\longrightarrow$ Seismic Design Category C
- $S_{DI} > 0.50$ $\longrightarrow$ Seismic Zone 4 $\longrightarrow$ Seismic Design Category D

2.9 WATER LOADING (WA)

Stream flow pressure exerted on piers in the longitudinal direction is determined by Equation 2.20. It is common practice to align substructure pier columns to coincide with the direction of flow, in which case the lateral pressure exerted on the pier columns is zero. In cases where the pier axis is not aligned with the direction of flow,

TABLE 2.17

Mississippi Embayment Site Factor F_v from Moon et al. (2016) – Site Class D

H, m	S_1	NEHRP	Upland	Lowland
30	0.10	2.40	2.20	2.30
30	0.20	2.00	1.87	1.93
30	0.30	1.80	1.64	1.63
30	0.40	1.60	1.45	1.38
30	0.50	1.50	1.34	1.24
100	0.10	2.40	3.18	3.12
100	0.20	2.00	2.59	2.58
100	0.30	1.80	2.17	2.33
100	0.40	1.60	1.81	1.86
100	0.50	1.50	1.61	1.60
200	0.10	2.40	3.60	3.52
200	0.20	2.00	3.08	3.11
200	0.30	1.80	2.56	2.66
200	0.40	1.60	2.19	2.07
200	0.50	1.50	1.94	1.77
300	0.10	2.40	3.78	3.76
300	0.20	2.00	3.29	3.36
300	0.30	1.80	2.73	2.85
300	0.40	1.60	2.29	2.15
300	0.50	1.50	2.02	1.83
500	0.10	2.40	3.81	3.79
500	0.20	2.00	3.32	3.39
500	0.30	1.80	2.76	2.88
500	0.40	1.60	2.32	2.18
500	0.50	1.50	2.05	1.86
1,000	0.10	2.40	3.83	3.81
1,000	0.20	2.00	3.34	3.41
1,000	0.30	1.80	2.78	2.90
1,000	0.40	1.60	2.34	2.20
1,000	0.50	1.50	2.07	1.88

a lateral pressure must also be accounted for, and is given by Equation 2.21. The stream flow velocity, V, must be in feet per second, and the resulting pressure, p, is in kips per square foot.

$$p = \frac{C_D V^2}{1,000} \tag{2.20}$$

$$p = \frac{C_L V^2}{1,000} \tag{2.21}$$

TABLE 2.18

Mississippi Embayment Site Factor F_v from Moon et al. (2016) – Site Class E

H, m	S_1	NEHRP	Upland	Lowland
30	0.10	3.50	3.00	3.06
30	0.20	3.20	2.58	2.36
30	0.30	2.80	2.15	1.78
30	0.40	2.40	1.85	1.33
30	0.50	2.40	1.64	1.08
100	0.10	3.50	3.90	4.40
100	0.20	3.20	3.38	3.48
100	0.30	2.80	2.58	2.77
100	0.40	2.40	2.27	2.26
100	0.50	2.40	1.95	1.82
200	0.10	3.50	4.90	4.80
200	0.20	3.20	4.06	3.74
200	0.30	2.80	3.14	3.17
200	0.40	2.40	2.82	2.63
200	0.50	2.40	2.32	2.13
300	0.10	3.50	5.26	5.06
300	0.20	3.20	4.11	3.93
300	0.30	2.80	3.14	3.35
300	0.40	2.40	2.82	2.68
300	0.50	2.40	2.42	2.33
500	0.10	3.50	5.26	5.50
500	0.20	3.20	4.21	4.40
500	0.30	2.80	3.14	3.48
500	0.40	2.40	2.84	2.83
500	0.50	2.40	2.45	2.54
1,000	0.10	3.50	5.26	5.50
1,000	0.20	3.20	4.21	4.52
1,000	0.30	2.80	3.14	3.65
1,000	0.40	2.40	2.84	2.98
1,000	0.50	2.40	2.47	2.65

The longitudinal drag coefficient, C_D, depends on the pier type and shape.

- Semi-circular nose pier, $C_D = 0.7$
- Square-ended pier, $C_D = 1.4$
- Debris lodged against pier, $C_D = 1.4$
- Wedge-nosed pier with nose angle 90 degrees or less, $C_D = 0.8$

The lateral drag coefficient, C_L, is a function of the angle, θ, between the flow direction and the longitudinal pier axis.

$\theta = 0$ degrees, $C_L = 0.0$
$\theta = 5$ degrees, $C_L = 0.5$
$\theta = 10$ degrees, $C_L = 0.7$
$\theta = 20$ degrees, $C_L = 0.9$
$\theta \geq 30$ degrees, $C_L = 1.0$

For additional guidance on stream flow pressure and other water-related loads, refer to Section 3.7 of the LRFD-BDS.

2.10 SOLVED PROBLEMS

Problem 2.1

Estimate the braking forces on each pier for a three-span bridge carrying two lanes of opposing traffic. However, traffic projections indicate that the bridge may be converted to a one-directional highway in the future with the construction of an adjacent, dual bridge. Span lengths are 168 ft, 240 ft, and 168 ft, for a total bridge length of 576 feet. Although the span arrangement is symmetrical, the pier heights are different: 28 feet at Pier 1 and 41 feet at Pier 2. For this example, assume that expansion joints exist at both abutments and the entire braking force is carried by the piers. Consider two trucks per loaded lane with the 10% reduction when such a condition is used for intermediate support reactions.

Assume that the bridge will be one-directional in the future and design for two lanes of braking forces with the corresponding multi-presence factor, $m = 1.0$. In addition, check the one-loaded-lane with multi-presence $m = 1.2$.

Problem 2.2

Consider a curved bridge with the radius of curvature of the traffic lane in question equal to $R = 730$ feet, superelevation, $SE = 0.08$ ft/ft, and design speed of 50 mph. The standard design truck with $h = b = 6$ feet will be used for the calculations. Determine the distribution of live load to the wheel lines and the centrifugal force, F_{CE}.

Problem 2.3

A 45 ft 3 inch wide, five-span bridge crosses a lake. Span lengths are 270 ft, 335 ft, 335 ft, 335 ft, and 270 ft for a total length of 1,545 ft. The bridge is located in an area with a mapped 3-second gust wind speed equal to 115 mph. The superstructure is 12.75 ft deep, measured from the top of the parapet to the bottom of the girder. Pier number 2 elevations are listed in the bullet points below. The pier consists of three 11 ft diameter columns. For the case of wind loading perpendicular to traffic, determine the design superstructure and substructure wind loads for Strength III, Strength V, Service I, and Service IV limit states.

- lake bed elevation = 520 ft
- normal pool elevation = 648 ft
- finished grade elevation = 733 ft

Problem 2.4

Suppose the Project Bridge is to be constructed in Martin, Tennessee on State Route TN-43 over State Route TN-22. Soil borings (hypothetical, not actual for Martin) indicate a subsurface profile as shown below. Determine the control points of the AASHTO design response spectrum, A_S, S_{DS}, S_{DI}, T_S, T_o, and T_L.

- Layer 1, Cohesionless, 65 ft thick, $N_i = 12$ blows/ft
- Layer 2, Cohesionless, 12 ft thick, $N_i = 18$ blows/ft
- Layer 3, Cohesive, 12 ft thick, $N_i = 27$ blows/ft
- Layer 4, Cohesionless, 16 ft thick, $N_i = 39$ blows/ft

Problem 2.5

A 1,545 ft long, steel I-girder bridge consists of five spans: 270 ft, 335 ft, 335 ft, 335 ft, and 270 ft for a total length of 1,545 ft. Both abutments have expansion joints. Successive pier heights are 47 ft, 47 ft, 93 ft, and 73 ft, beginning at Pier No. 1. Identical column cross sections are used at each pier and the bridge is located in a cold climate. Determine (a) the expansion joint requirements and (b) the movement induced on each pier due to uniform temperature change (TU) effects.

Problem 2.6

Determine the shear and moment at tenth-points of an interior girder for the steel girder option of the Project Bridge for the following loads. Use a flange width equal to 16 inches for concrete in the haunch. Take the girder spacing, $S = 9$ ft 3 inches and use a W40 × 215 girder. Include 7.5% of the girder self-weight in DC1 to account for miscellaneous items. Each parapet weighs 0.400 klf.

- Dead load on non-composite section (DC1)
- Dead load on composite section (DC2)
- Dead load wearing surface (DW)

PROBLEM 2.1	TEH	1/1

Case 1. $0.90 \times (25\%$ of two design trucks in each lane) with $m = 1.0$.

$$F_{BR} = 0.90\left[0.25\left(72k \times 2\,\text{trucks}\right)\right]\left(2\,\text{lanes}\right) = 64.8\,\text{kips}$$

Case 2. $0.90 \times (5\%$ of two design trucks in each lane plus uniform lane load) with $m = 1.0$.

$$F_{BR} = 0.90\left[0.05\left(0.640\,\text{klf} \times 576\,\text{ft} + 72k \times 2\,\text{trucks}\right)\right]\left(2\,\text{lanes}\right) = 46.1\,\text{kips}$$

Case 3. $0.90 \times (5\%$ of two design trucks in one lane plus uniform lane load) with $m = 1.2$.

$$F_{BR} = 0.90\left[0.05\left(0.640\,\text{klf} \times 576\,\text{ft} + 72k \times 2\,\text{trucks}\right)\right]\left(1\,\text{lane}\right)\left(1.2\right) = 27.7\,\text{kips}$$

Case 4. $0.90 \times (25\%$ of two design trucks in one lane) with $m = 1.2$.

$$F_{BR} = 0.90 \left[0.25 \left(72\text{k} \times 2\,\text{trucks} \right) \right] (1\,\text{lane})(1.2) = 38.9\,\text{kips}$$

Case 1 controls and the total braking force of 64.8 kips can be distributed to the piers. Set the relative stiffness, $K_R = 1.0$ for Pier 2, the tallest pier. Recognize that elastic stiffness is proportional to the inverse of the height cubed.

$$K_{R1} = 1.000 \left(\frac{41}{28} \right)^3 = 3.140$$

$$F_{BR-P1} = 64.8 \left(\frac{3.140}{3.140 + 1.000} \right) = 49.1\,\text{kips}$$

$$F_{BR-P2} = 64.8 \left(\frac{1.000}{3.140 + 1.000} \right) = 15.7\,\text{kips}$$

PROBLEM 2.2	TEH	1/1

After converting the design speed to $v = 73.33$ feet per second, find $C = 0.305$ (with $f = 4/3$).

For $SE = 0.08$, $\phi = 0.07983$ radians. Solving the equations in the text gives the following:

$F_{N2}/W = 0.735$ (the outside wheel normal force is 73.5% of the truck weight)

$F_{N1}/W = 0.286$ (the inside wheel normal force is 28.6% of the truck weight)

$F_{T1}/W = 0.063$ (the inside wheel traction force is 6.3% of the truck weight)

$F_{T2}/W = 0.161$ (the outside wheel traction force is 16.1% of the truck weight)

$F_{V2}/W = 0.720$ (the outside wheel vertical force is 72.0% of the truck weight)

$F_{V1}/W = 0.280$ (the inside wheel vertical force is 28.0% of the truck weight)

$$F_{CE} = 0.305 \times 72 = 22.0 \text{ kips per lane}$$

PROBLEM 2.3	TEH	1/4

Pier 2 supports one-half of
span 2 & one-half of span 3.
$\rightarrow$ Tributary Length = $\frac{1}{2}(335+335)$
$= 335'$

Strength III, $V = 115\ mph$
Strength V, $V = 80\ mph$
Service I, $V = 70\ mph$
Service IV, $V = .75 \times 115 = 86.3\ mph$

$G = 1.0 \qquad C_d = 1.3,$ superstructure
$\qquad\qquad\qquad 1.6,$ substructure

Strength III & Service IV:
$\qquad K_z$ by Equations
Strength V & Service I:
$\qquad K_z = 1.00$

Vertical Pressure at Deck $1/4$-pt.
$\qquad 20\ psf,$ Strength III
$\qquad 10\ psf,$ Service IV
$\qquad 0\ psf,$ Strength V &
$\qquad\qquad$ Service I

PROBLEM 2.3	TEH	2/4

Strength III

$$Z = 733 - 648 = 85'$$

Design for $Z = 100'$ to accomodate water elevation below normal pool.

Use Exposure Category "D"

$$K_z(D) = \frac{\left[2.5 \ln \left(\frac{100}{.0164} \right) + 7.65 \right]^2}{616.1}$$

$$\Rightarrow K_z(D) = 1.407$$

$$P_z = 2.56 \times 10^{-6} (115)^2 (1.407)(1.00) C_D$$

$$= 0.0476 \, C_D$$

$$= 0.0476 \, (1.3) = 0.062 \, KSF \\ (\text{superstructure})$$

$$= 0.0476 \, (1.6) = 0.076 \, KSF \\ (\text{substructure})$$

$$(F_H)_{WS} = 0.062 \times 335' \times 12.75' = 265 \, KIPS$$

$$(F_V)_{WS} = 0.020 \, KSF \times 335' \times 45.25' = 303 \, KIPS$$

$$(w)_{WS} = 0.076 \times 11' = 0.836 \, KLF$$

$(F_H)_{WS}$: horizontal load at Pier top
$(F_V)_{WS}$: vertical load at Pier top
$(w)_{WS}$: uniform lateral load on column

PROBLEM 2.3	TEH	3/4

Strength $\underline{V}$

$P_Z = 2.56 \times 10^{-6} (80)^2 (1.00)(1.0) C_D$

$\quad = 0.0164 \, C_D$

$\quad = 0.0164 \, (1.3) = 0.0213 \, \text{KSF}$
$\qquad\qquad\qquad (\text{superstructure})$

$\quad = 0.0164 \, (1.6) = 0.0262 \, \text{KSF}$
$\qquad\qquad\qquad (\text{substructure})$

$(F_H)_{WS} = 0.0213 \times 335' \times 12.75' = 91 \, \text{KIPS}$

$(F_r)_{WS} = 0$

$(w)_{WS} = 0.0262 \times 11' = 0.288 \, \text{KLF}$

Service $\underline{I}$

$P_Z = 2.56 \times 10^{-6} (70)^2 (1.00)(1.00) C_D$

$\quad = 0.0125 \, C_D$

$\quad = 0.0125 \, (1.3) = 0.0163 \, \text{KSF}$
$\qquad\qquad\qquad (\text{superstructure})$

$\quad = 0.0125 \, (1.6) = 0.020 \, \text{KSF}$
$\qquad\qquad\qquad (\text{substructure})$

$(F_H)_{WS} = 0.0163 \times 335' \times 12.75' = 70 \, \text{KIPS}$

$(F_r)_{WS} = 0$

$(w)_{WS} = 0.020 \times 11' = 0.220 \, \text{KLF}$

PROBLEM 2.3	TEH	4/4

$\underline{Service\ IV}$

$$P_Z = 2.56 \times 10^{-6} (86.3)^2 (1.00)(1.407) C_D$$

$$= 0.0268\ C_D$$

$$= 0.0268\ (1.3) = 0.0349\ \text{KSF}$$

$$(superstructure)$$

$$= 0.0268\ (1.6) = 0.0429\ \text{KSF}$$

$$(substructure)$$

$$(F_H)_{ws} = 0.0349 \times 335' \times 12.75' = 149\ \text{KIPS}$$

$$(F_V)_{ws} = 0.010 \times 335' \times 45.25' = 152\ \text{KIPS}$$

$$(w)_{ws} = 0.0429 \times 11' = 0.472\ \text{KLF}$$

PROBLEM 2.4	TEH	1/4

Based on all four layers, determine the average blow count.

$$(N)_1 = \frac{\Sigma d_i}{\Sigma \dfrac{d_i}{N_i}} = \frac{100}{\dfrac{65}{12} + \dfrac{12}{18} + \dfrac{7}{27} + \dfrac{16}{39}} = 14.8 < 15 \rightarrow \text{Site Class E}$$

Using only the cohesionless layers, determine the average blow count.

$$(N)_2 = \frac{\Sigma d_i}{\Sigma \dfrac{d_i}{N_i}} = \frac{93}{\dfrac{65}{12} + \dfrac{12}{18} + \dfrac{16}{39}} = 14.3 < 15 \rightarrow \text{Site Class E}$$

Use Site Class E site factors, F_{PGA}, F_a, and F_v.

Use the Unified Hazard Tool (UHT) at the United States Geological Survey (USGS) web site (https://earthquake.usgs.gov/hazards/interactive/) and determine the 7% in 75-year probability of exceedance bedrock PSA values for Martin, Tennessee.

$$MRI = \frac{-t}{\ln(1-PE)} = \frac{-75}{\ln(1-0.07)} = 1{,}033\ \text{years}$$

The UHT online application may also be used to pinpoint the project coordinates.

Obtain PSA at 0 seconds (PGA), at 0.2 seconds (SS), and at 1.0 second (S1) at the B/C boundary.

Though not required for bridges to be designed using a dynamic response spectrum analysis, hazard de-aggregation may also be obtained at the site to reveal magnitude and distance pairs characterizing the hazard at the site. Such de-aggregation will prove essential when designing using response history analysis on computer models subject to ground motion accelerograms.

PROBLEM 2.4	TEH	2/4

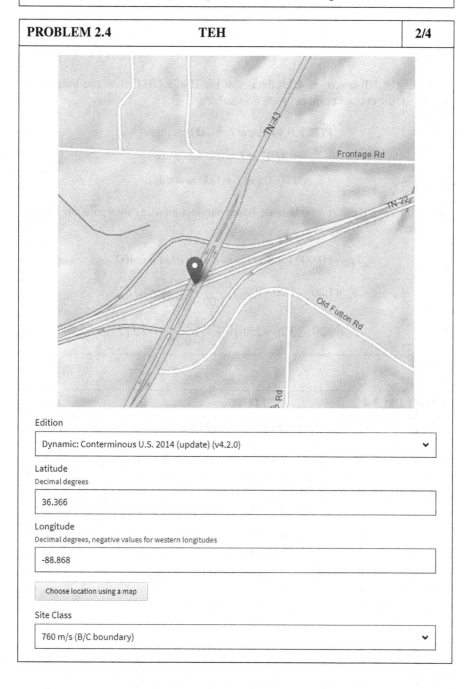

Edition

Dynamic: Conterminous U.S. 2014 (update) (v4.2.0)	⌄

Latitude
Decimal degrees

36.366

Longitude
Decimal degrees, negative values for western longitudes

-88.868

Choose location using a map

Site Class

760 m/s (B/C boundary)	⌄

PROBLEM 2.4	TEH	3/4

Spectral Period

Peak Ground Acceleration   ⌄

Time Horizon
Return period in years

1033

Using the 2014, version 4.2.0 dataset at the USGS UHT gives the following values for the control point bedrock *PSA* values:

$$PGA = 0.4472 \ g \text{ at } T = 0.00 \text{ seconds}$$

$$S_S = 0.7607 \text{ at } T = 0.20 \text{ seconds}$$

$$S_1 = 0.2070 \ g \text{ at } T = 1.00 \text{ seconds}$$

From Tables in the text, obtain code-based site factors using interpolation where required.

$$F_{PGA} = 0.900 \text{ for Site Class } E \text{ with } PGA > 0.400$$

$$F_a = 1.20 - (1.20 - 0.90)\frac{0.7607 - 0.75}{1.00 - 0.75} = 1.187$$

$$F_v = 3.20 - (3.20 - 2.80)\frac{0.2070 - 0.20}{0.30 - 0.20} = 3.172$$

PROBLEM 2.4	TEH	4/4

The surface, design response spectrum control points may now be determined.

$$A_s = 0.4472 \times 0.900 = 0.4025$$

$$S_{DS} = 0.7607 \times 1.187 = 0.9029$$

$$S_{D1} = 0.2070 \times 3.172 = 0.6566$$

$$T_S = \frac{0.6566}{0.9029} = 0.727 \text{ seconds}$$

$$T_o = 0.20 \times 0.727 = 0.145 \text{ seconds}$$

For this problem, de-aggregation reveals a modal M, $R = 7.76$, 50 km for all three control points. The modal M, R combination is the one most likely to produce ground motion exceeding the design value.

The ATC Hazard by Location tool may be used to find that the transition period, T_L, = 12 seconds for the project site.

PROBLEM 2.5	TEH	1/3

steel, $\alpha = 0.0000065 /F^\circ$

Cold climate, $\delta T = -30 - 120$

$\qquad\qquad\qquad = 150 F^\circ$

Pier $K_i \propto \dfrac{1}{H_i^3}$

Take $K_3 = 1.00$ $(H = 93')$

$\qquad K_1 = K_2 = 1.00 \left(\dfrac{93}{47}\right)^3 = 7.747$

$\qquad K_4 = 1.00 \left(\dfrac{93}{73}\right)^3 = 2.068$

$\qquad\qquad \Rightarrow \sum_i K_i = 18.562$

$X_{P1} = 270'$ $\qquad X_{P2} = 270 + 335 = 605'$

$X_{P3} = 605 + 335 = 940'$

$X_{P4} = 940 + 335 = 1,275'$

$X_{A2} = 1,275 + 270 = 1,545'$ ✓

$L_{A1} = \dfrac{7.747(270 + 605) + 1.0(940) + 2.068(1,275)}{18.562}$

$\qquad\qquad \Rightarrow L_{A1} = 558'$ from Abutment 1
$\qquad\qquad\qquad\qquad$ to center of
$\qquad\qquad\qquad\qquad$ stiffness

$L_{P1} = 558 - 270 = 288'$

$L_{P2} = 605 - 558 = 47'$

PROBLEM 2.5	TEH	2/3

$$L_{P3} = 940 - 558 = 382'$$
$$L_{P4} = 1,275 - 558 = 717'$$
$$L_{A2} = 1,545 - 558 = 987'$$

For each 100' of distance:
$$\Delta_{TU} = (100 \times 12)(150 F°)(0.0000065)$$
$$= 1.17'' \text{ per } 100\text{-ft}$$

Total Movements:
$$(\Delta_{TU})_{A1} = 1.17(558/100) = 6.53''$$
$$\underline{\times \quad 1.20}$$
$$\overline{7.83''} \longrightarrow$$

$$(\Delta_{TU})_{A2} = 1.17(987/100)(1.20)$$
$$\underline{= 13.86''}$$

Provide expansion joints & bearings at both abutments:

Abutment 1: 8'' total
                    (4'' exp. /4'' contr.)

Abutment 2: 14'' total
                    (7'' exp. /7'' contr.)

PROBLEM 2.5	TEH	3/3

$$(\Delta_{TU})_{P1} = 1.17(288/100)(1.20)$$

$$= 4.04'' \text{ total}$$

$$= 2.02'' \text{ each way}$$

$$(\Delta_{TU})_{P2} = 1.17(47/100)(1.20)$$

$$= 0.66'' \text{ total}$$

$$= 0.33'' \text{ each way}$$

$$(\Delta_{TU})_{P3} = 1.17(382/100)(1.20)$$

$$= 5.36'' \text{ total}$$

$$= 2.68'' \text{ each way}$$

$$(\Delta_{TU})_{P4} = 1.17(717/100)(1.20)$$

$$= 10.07'' \text{ total}$$

$$= 5.03'' \text{ each way}$$

With fixed bearings at all Piers, the required movements must be accomodated thru flexure of the Pier Columns.

PROBLEM 2.6	TEH	1/4

Deck weight $= (8.25/12)(9.25)(0.150) = 0.954$ klf per interior girder
Haunch weight $= (2/12)(16/12)(0.150) = 0.033$ klf per interior girder
Girder weight $= 0.215 \times 1.075 = 0.226$ klf per interior girder

- DC1 $= 0.954 + 0.033 + 0.226 = 1.213$ klf per interior girder

Parapet weight $= 2 \times 0.400/4$ girders $= 0.200$ klf per interior girder

- DC2 $= 0.200$ klf per interior girder

Overlay = 32 ft × 0.035 ksf/4 girders = 0.280 klf per interior girder

- DW = 0.280 klf per interior girder

Use VisualAnalysis to create a two-span, continuous beam model:

CEE4380-Problem-02-06.vap

Notes:

- It is common practice to assume that DC1 weight applied to a girder includes the self-weight of the girder plus the tributary width of the deck slab.
- Overlay load is commonly distributed equally to all girders.
- Some engineers distribute parapet loads equally to all girders, but it is not uncommon to find offices in which parapet loads are assumed to be carried only by the exterior and adjacent interior girders.

PROBLEM 2.6	TEH						2/4

Due to symmetry, it is only necessary to print results for one span.

Member	Result Case	Offset	Fx	Vy	Vz	Torsion	My	Mz
		ft	K	K	K	K-ft	K-ft	K-ft
BmX001	DC1	0	0	40.9	0	0	0	0.0
BmX001	DC1	9	0	30.0	0	0	0	319.3
BmX001	DC1	18	0	19.1	0	0	0	540.4
BmX001	DC1	27	0	8.2	0	0	0	663.2
BmX001	DC1	36	0	-2.7	0	0	0	687.8
BmX001	DC1	45	0	-13.6	0	0	0	614.1
BmX001	DC1	54	0	-24.6	0	0	0	442.1
BmX001	DC1	63	0	-35.5	0	0	0	171.9
BmX001	DC1	72	0	-46.4	0	0	0	-196.5
BmX001	DC1	81	0	-57.3	0	0	0	-663.2
BmX001	DC1	90	0	-68.2	0	0	0	-1,228.2
BmX001	DC2	0	0	6.8	0	0	0	0.0
BmX001	DC2	9	0	5.0	0	0	0	52.7
BmX001	DC2	18	0	3.2	0	0	0	89.1
BmX001	DC2	27	0	1.4	0	0	0	109.4
BmX001	DC2	36	0	-0.5	0	0	0	113.4
BmX001	DC2	45	0	-2.3	0	0	0	101.3
BmX001	DC2	54	0	-4.1	0	0	0	72.9
BmX001	DC2	63	0	-5.9	0	0	0	28.4
BmX001	DC2	72	0	-7.7	0	0	0	-32.4
BmX001	DC2	81	0	-9.5	0	0	0	-109.4
BmX001	DC2	90	0	-11.3	0	0	0	-202.5
BmX001	DW	0	0	9.5	0	0	0	0.0
BmX001	DW	9	0	6.9	0	0	0	73.7
BmX001	DW	18	0	4.4	0	0	0	124.7
BmX001	DW	27	0	1.9	0	0	0	153.1
BmX001	DW	36	0	-0.6	0	0	0	158.8
BmX001	DW	45	0	-3.2	0	0	0	141.8
BmX001	DW	54	0	-5.7	0	0	0	102.1
BmX001	DW	63	0	-8.2	0	0	0	39.7
BmX001	DW	72	0	-10.7	0	0	0	-45.4
BmX001	DW	81	0	-13.2	0	0	0	-153.1
BmX001	DW	90	0	-15.8	0	0	0	-283.5

PROBLEM 2.6	TEH	3/4

DC1

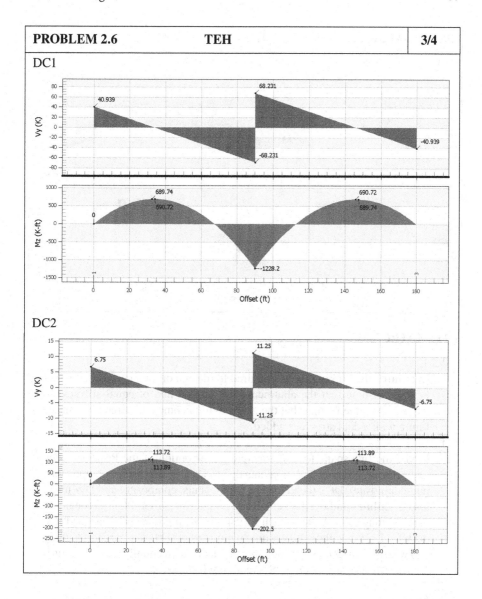

DC2

PROBLEM 2.6	TEH	4/4

DW

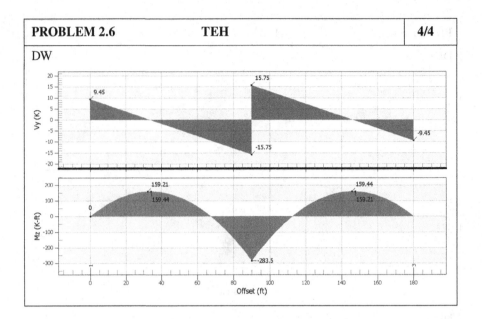

2.11   EXERCISES

E2.1.

Determine the braking forces required for design of the pier for the Project Bridge. Assume that the abutments are integral and of equal stiffness to the pier. Assume further that the bridge will become one-directional in the future with a dual structure added.

E2.2.

Model the two spans for the concrete girder option of the Project Bridge and determine shears and moments due to DC1 (dead load of components, non-composite), DC2 (dead load of components, composite) and DW (dead load of utilities and overlay) effects on an interior girder. Plot the shear and moment diagrams. Assume each BT-54 girder supports a tributary width of deck for DC1 calculations. Assume parapet (0.400 klf per parapet) loading and wearing surface (35 psf) loading are distributed equally to all four girders. Take the girder spacing, 'S' equal to 9 ft 3 inches.

E2.3.

Suppose the Project Bridge is to be constructed in Union City, Tennessee on State Route TN-22 over N. Clover Street. Soil borings (hypothetical, not actual for Union City) indicate a subsurface profile as shown in Table E2.3 below. Determine the control points of the AASHTO design response spectrum, A_S, S_{DS}, S_{D1}, T_S, T_o, and T_L. Consider two cases: (a) using AASHTO site factors and (2) using the Hashash et al. (2008) uplands site factors. Plot the surface spectra for each case and compare. Use an estimated subsurface profile depth of 500 meters.

TABLE E2.3
Soil Profile for Exercise E2.3

Layer	Thickness, d_i (feet)	Blow Count, N_i (blows/ft)
1, cohesionless	5	7
2, cohesionless	5	8
3, cohesive	13	36
4, cohesionless	10	17
5, cohesionless	10	24
6, cohesionless	10	44
7, cohesionless	10	51
8, cohesionless	10	57
9, cohesive	17	69
10, cohesionless	10	32

E2.4.

A continuous steel girder bridge consists of 234-ft end spans with five 300-ft interior spans for a total bridge length of 1,968 feet. Pier heights are 47 feet, 47 feet, 73 feet, 73 feet, 31 feet, and 17 feet. Expansion joints and bearings are used at both abutments only. Pier cross-section geometry is identical at all piers. The bridge is in a cold climate. Estimate the expansion joint requirements at Abutment 1 and Abutment 2.

E2.5.

Figure E2.5 is a cross section, looking forward, of a curved bridge with centerline radius equal to 800 feet. The bridge curves to the right. Determine the centrifugal force on the pier for the two-span bridge. Spans are equal and the total bridge length is 280 feet. The design speed is 40 mph, and the bridge is one-directional.

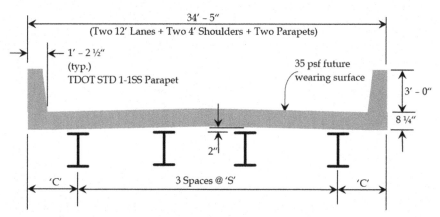

FIGURE E2.5

E2.6.

A bridge in Jackson, Tennessee is considered critical and will be designed for the 2,500-year ground shaking rather than the standard 1,033-year ground shaking. $S_S=0.7905$ and $S_I=0.2457$ for the 2500-year MRI. The site is an uplands Class D site. Assess the various options for site factors and develop a proposed design response spectrum for the project.

3 Load Combinations and Limit States

Chapter 2 presents a discussion of the basic load and resistance factor design (LRFD) relationship. This discussion includes the idea of both load (γ) factors and resistance (ϕ) factors, which will be further elaborated on in the current chapter.

The American Association of State Highway and Transportation Officials (AASHTO) LRFD Bridge Design Specifications (BDS) design philosophy incorporates various limit states into design criteria. The limit states include:

- Strength I limit state: normal traffic with no wind load
- Strength II limit state: a means for owners to specify special design vehicles
- Strength III limit state: maximum design wind load on the bridge with no traffic
- Strength IV limit state: appropriate for bridges dominated by dead load effects
- Strength V limit state: an intermediate condition between Strength I and Strength III
- Service I limit state: normal operation with a wind velocity equal to 70 mph
- Service II limit state: control of yielding in steel and slip in slip-critical connections
- Service III limit state: tension in prestressed concrete superstructures
- Service IV limit state: crack control in prestressed concrete columns
- Fatigue I limit state: infinite life, load-induced fatigue
- Fatigue II limit state: finite life, load-induced fatigue
- Extreme Event I limit state: earthquake loading with collapse prevention
- Extreme Event II limit state: blast, ice, vehicle collision, and vessel collision

For a typical I-girder bridge, it would be necessary to evaluate Strength I, III, and V limit states. Strength IV evaluation is necessary for bridges with unusually high dead load-to-live load ratio. Service limit states applicable for precast, prestressed concrete I-girder bridges include Service I and Service III. For steel I-girder bridges, Service limit states I and II apply.

Load sources at the various limit states in AASHTO are characterized as either permanent or transient, and are defined as follows:

Permanent Loads

- CR = force effects due to creep
- DD = downdrag force

DOI: 10.1201/9781003265467-3

- DC = dead load of structural components and nonstructural attachments
- DW = dead load of wearing surfaces and utilities
- EH = horizontal earth pressure load
- EL = miscellaneous locked-in force effects from the construction process
- ES = earth surcharge load
- EV = vertical pressure from dead load of earth fill
- PS = total prestress forces for Service limit states
- SH = force effects due to shrinkage

Transient Loads

- BL = blast loading
- BR = vehicular braking force
- CE = vehicular centrifugal force
- CT = vehicular collision force
- CV = vessel collision force
- EQ = earthquake load
- FR = friction load
- IC = ice load
- IM = vehicular dynamic load allowance
- LL = vehicular live load
- LS = live load surcharge
- PL = pedestrian live load
- SE = force effect due to settlement
- TG = force effect due to temperature gradient
- TU = force effect due to uniform temperature
- WA = water load and stream pressure
- WL = wind on live load
- WS = wind load on structure

Equation 3.1 is the basic LRFD-based design equation, with ductility (η_D), redundancy (η_R), and importance (η_I) modifiers all equal to 1.0. This will be the case for most routine bridge designs. Refer to Chapter 2 for additional details on the load modifiers, and situations when values equal to 1.0 may not be appropriate.

$$Q_n = \Sigma \gamma_i Q_i \leq \phi R_n \tag{3.1}$$

Tables 3.1 through 3.4 summarize load factors (γ) for the specified limit states. Note that, for many limit states, multiple factors are listed for TU-loading (uniform temperature variation). The smaller value is intended to be used for force-related actions, and the larger value for deformation-related actions.

Table 3.5 lists the variable load factors, γ_p, for the Strength limit state load combinations. For variable load factors related to thermal gradient (TG) and support settlement (SE), the reader is referred to Chapter 3 of the LRFD-BDS.

For the Extreme Event II limit state, only one extreme load at a time (BL, IC, CT, or CV) is considered.

TABLE 3.1
Strength Limit State Load Factors

Limit State	DC DD DW EH EV ES EL PS CR SH	LL IM CE BR PL LS	WA	WS	WL	FR	TU	TG	SE
Strength I	γ_p	1.75	1.00	–	–	1.00	0.50/1.20	γ_{TG}	γ_{SE}
Strength II	γ_p	1.35	1.00	–	–	1.00	0.50/1.20	γ_{TG}	γ_{SE}
Strength III	γ_p	–	1.00	1.00	–	1.00	0.50/1.20	γ_{TG}	γ_{SE}
Strength IV	γ_p	–	1.00	–	–	1.00	0.50/1.20	–	–
Strength V	γ_p	1.35	1.00	1.00	1.00	1.00	0.50/1.20	γ_{TG}	γ_{SE}

TABLE 3.2
Service Limit State Load Combinations

Limit State	DC DD DW EH EV ES EL PS CR SH	LL IM CE BR PL LS	WA	WS	WL	FR	TU	TG	SE
Service I	1.00	1.00	1.00	1.00	1.00	1.00	1.00/1.20	γ_{TG}	γ_{SE}
Service II	1.00	1.30	1.00	–	–	1.00	1.00/1.20	–	–
Service III	1.00	γ_{LL}	1.00	–	–	1.00	1.00/1.20	γ_{TG}	γ_{SE}
Service IV	1.00	–	1.00	1.00	–	1.00	1.00/1.20	–	1.00

A subset of LRFD-BDS resistance factors for steel elements and concrete elements is summarized in Tables 3.6 and 3.7.

Exceptions to Tables 3.6 and 3.7 are made at the Extreme Event limit states. At the Extreme Event limit state, resistance factors are 1.00 for failure modes other than those shown for ASTM F 3125 bolts and shear connectors.

For concrete elements, it is necessary to determine whether tension-control or compression-control governs the resistance factor.

TABLE 3.3
Extreme Event Limit State Load Combinations

Limit State	DC DD DW EH EV ES EL PS CR SH	LL IM CE BR PL LS	WA	FR	EQ	BL	IC	CT	CV
Extreme Event I	1.00	γ_{EQ}	1.00	1.00	1.00	–	–	–	–
Extreme Event II	1.00	0.50	1.00	1.00	–	1.00	1.00	1.00	1.00

TABLE 3.4
Fatigue Limit State Load Combinations

Limit State	LL IM CE
Fatigue I	1.75
Fatigue II	0.80

TABLE 3.5
Variable Load Factors

Load and Limit State	Maximum γ_p	Minimum γ_p
DC – Strength I, II, III, V	1.25	0.90
DC – Strength IV	1.50	0.90
DW	1.50	0.65
EH – Active	1.50	0.90
EH – At Rest	1.35	0.90
EV – Overall Stability	1.00	NA
EV – Retaining Walls & Abutments	1.35	1.00
ES	1.50	0.75
PPC with Refined Losses/Elastic Gains	$\gamma_{LL}=1.00$	
All other PPC	$\gamma_{LL}=0.80$	

TABLE 3.6

Resistance Factors for Steel Elements

Flexure	$\phi_f = 1.00$
Shear	$\phi_v = 1.00$
Axial compression, steel only	$\phi_c = 0.95$
Flexure with axial compression in CFSTs	$\phi_c = 0.90$
Axial compression in composite columns	$\phi_c = 0.90$
Net section tensile fracture	$\phi_u = 0.80$
Gross section tensile yielding	$\phi_y = 0.95$
Bearing on milled surfaces	$\phi_b = 1.00$
Bolts bearing on material	$\phi_{bb} = 0.80$
Shear connectors	$\phi_{sc} = 0.85$
ASTM F 3125 bolts in tension	$\phi_t = 0.80$
ASTM F 3125 bolts in shear	$\phi_s = 0.80$
ASTM F1554 anchor rod in tension	$\phi_t = 0.80$
ASTM F1554 anchor rod in shear	$\phi_s = 0.75$
Block shear	$\phi_{bs} = 0.80$
Shear rupture in connected elements	$\phi_{vu} = 0.80$
Web crippling	$\phi_w = 0.80$
Weld metal, complete penetration welds	$\phi_{e1} = 0.85$
Weld metal, fillet welds	$\phi_{e2} = 0.80$
Resistance during pile driving	$\phi = 1.00$

TABLE 3.7

Resistance Factors for Normal Weight Concrete Elements

Tension-controlled reinforced concrete sections	$\phi = 0.90$
Tension-controlled prestressed concrete with bonded strand	$\phi = 1.00$
Shear and torsion in reinforced concrete	$\phi = 0.90$
Compression-controlled sections with spirals or ties	$\phi = 0.75$
Bearing on concrete	$\phi = 0.70$
Resistance during pile driving	$\phi = 1.00$

Tension-controlled reinforced concrete sections are defined as sections with a net tensile strain in the extreme layer of tensile reinforcement, ε_t, greater than or equal to the tension-controlled strain limit, ε_{tl}, when the concrete strain reaches a value of 0.003.

Compression-controlled reinforced concrete sections are defined as sections with a net tensile strain in the extreme layer of tensile reinforcement, ε_t, less than or equal to the compression-controlled strain limit, ε_{cl}, when the concrete strain reaches a value of 0.003.

The tension-controlled strain limit, ε_{tl}, is determined as follows:

- $\varepsilon_{tl} = 0.005$ for reinforcement with $f_y \leq 75$ ksi
- $\varepsilon_{tl} = 0.008$ for reinforcement with $f_y = 100$ ksi
- ε_{tl} is determined by linear interpolation for reinforcement with $75 < f_y < 100$ ksi

The compression-controlled strain limit, ε_{cl}, is determined as follows.

- $\varepsilon_{cl} = f_y/E_s$, but not > 0.002, for reinforcement with $f_y \leq 60$ ksi
- $\varepsilon_{cl} = 0.004$ for reinforcement with $f_y = 100$ ksi
- ε_{cl} is determined by linear interpolation for reinforcement with $60 < f_y < 100$ ksi

For sections with net tensile strain in the extreme layer of tension reinforcement between ε_{cl} and ε_{tl}, linear interpolation, as given below in Equation 3.2, is used to determine the appropriate resistance factor:

$$0.75 \leq \phi = 0.75 + \frac{0.15\left(\varepsilon_t - \varepsilon_{cl}\right)}{\varepsilon_{tl} - \varepsilon_{cl}} \leq 0.90 \tag{3.2}$$

For concrete stress-strain profiles, the AASHTO LRDS-BDS permits the use of "rectangular, parabolic, or any other shape that results in a prediction of strength in substantial agreement with test results" (Section 5.6.2). For hand calculations, the traditional rectangular assumption is often used. For section analysis by computer, with Response 2000, for example, a parabolic assumption is not uncommon.

3.1 SOLVED PROBLEMS

Problem 3.1

Figure P3.1 represents a single girder line of a three-span continuous bridge. The reactions for a single girder at abutment number 1 and pier number 1 are summarized in Table P3.1. Determine the Strength limit state girder reactions at abutment number 1 and pier number 1. Negative reactions indicate uplift.

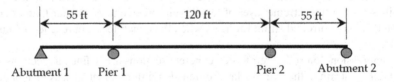

FIGURE P3.1 Problem 3.1.

TABLE P3.1
Reactions for Problem P3.1

Load Case	Abutment 1 Reaction	Pier 1 Reaction
DC	15.1 kips	174.6 kips
DW	2.6 kips	29.6 kips
LL+IM (Maximum)	72.3 kips	143.8 kips
LL+IM (Minimum)	−26.6 kips	−5.7 kips

Problem 3.2

Section 6.6.1.2. of the AASHTO LRFD-BDS requires that fatigue be investigated whenever unfactored compressive dead load stress is less than Fatigue I limit state tensile stress for the detail under consideration. At a particular point in a plate girder, long-term composite properties are used for composite dead load stress calculations (DC2 and DW), girder properties alone are used for non-composite dead load effects (DC1), and short-term composite properties are used for live load and fatigue stress calculations. Determine whether flexural fatigue stress range calculations need to be made for the top flange and for the bottom flange. Properties (section moduli, Table P3.2a) and unfactored moments (Table P3.2b) are provided.

Positive moment causes compression in the top flange, tension in the bottom flange.

TABLE P3.2A
Section Properties for Problem P3.2

Section	S_{top}, in^3	S_{bott}, in^3
Girder alone	1,088	1,088
Short-term composite	14,261	1,470
Long-term composite	3,867	1,359

TABLE P3.2B
Moments for Problem P3.2

Load Case	Moment, ft-kips
DC1	181
DC2	29
DW	40
Fatigue + IM (Maximum)	296
Fatigue + IM (Minimum)	−175

Problem 3.3

The top of a circular, cantilever pier column supporting a curved bridge is subjected to the actions summarized in Table P3.3. The single-post column height, H, is 82 feet from the point of load application to fixity at the base. Determine Strength I, Strength III, Strength V, and Service I axial load,

TABLE P3.3

Column Loads for Problem 3.3

Load Case	P, kips	V_T, kips	V_L, kips	M_T, ft-k	M_L, ft-k
DC	1,922	–	–	–	89
DW	272	–	–	–	–
LL+IM	744	–	–	–	3,929
CE	–	33	–	–	396
BR	–	–	36	432	–
WS	–	89	22	–	801
WL	–	18	8	–	324
TU	–	23	30	–	–

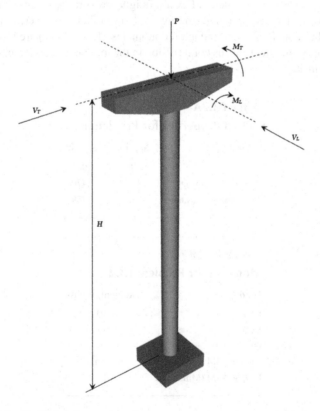

FIGURE P3.3   Problem 3.3.

moment, and shear acting on the base of the column. Values for WS in the table are for the Strength III limit state wind speed of 115 mph. Values for other limit states should be adjusted accordingly (Figure P3.3):

- P = axial compression
- V_T = shear in the transverse direction, perpendicular to traffic
- V_L = shear in the longitudinal direction, parallel to traffic
- M_T = moment about a transverse axis
- M_L = moment about a longitudinal axis

Problem 3.4

A 48-inch square reinforced concrete bridge column consists of 20 #10 bars distributed uniformly around the perimeter. Clear cover is 2.5 inches. Concrete and steel strengths are f'_c = 4,000 psi and f_y = 60 ksi. The moment-to-shear ratio for the column is 20 feet for the limit state under investigation. Determine the appropriate resistance factor, ϕ, and the design resistance, ϕM_n, for the following cases:

a) Axial load, P_u = 0
b) Axial load, P_u = 2,230 kips compression

PROBLEM 3.1	TEH	1/1

Strength I Limit state.

Abutment No. 1. (Maximum)

$$R_{A1} = 1.25(15.1) + 1.5(2.6)$$
$$+ 1.75(72.3)$$

$$\Rightarrow \underline{R_{A1} = 149.3 \text{ KIPS}}$$

Abutment No. 1. (Minimum)

$$R_{A1} = 0.9(15.1) + 0.0(2.6)$$
$$+ 1.75(-26.6)$$

$$\Rightarrow \underline{R_{A1} = -33.0 \text{ KIPS}}$$

Pier No. 1. (Maximum)

$$R_{P1} = 1.25(174.6) + 1.50(29.6)$$
$$+ 1.75(143.8)$$

$$\Rightarrow \underline{R_{P1} = 514.3 \text{ KIPS}}$$

Pier No. 1 . (Minimum)

$$R_{P1} = 0.9(174.6) + 0.0(29.6)$$
$$+ 1.75(-5.7)$$

$$\Rightarrow R_{P1} = 147.2 \text{ KIPS}$$

(Uplift possible at Abutment)

PROBLEM 3.2	**TEH**	**1/1**

Top Flange:

$$f = \frac{181 \times 12}{1,088} + \frac{29 \times 12}{3,867} + \frac{40 \times 12}{3,867}$$
$$+ \frac{1.75(-175 \times 12)}{14,261}$$

$$\Rightarrow f = +1.95 \text{ ksi}$$
(no net tension; no
Fatigue assessment for
top flange)

Bottom Flange:

$$f = \frac{-181 \times 12}{1,088} - \frac{29 \times 12}{1,359} - \frac{40 \times 12}{1,359}$$
$$- \frac{1.75(296 \times 12)}{1,470}$$

$$\Rightarrow f = -12.83 \text{ ksi, net tension}$$

Bottom flange Fatigue I limit state:

$$\gamma \Delta f = \frac{1.75(296 - (-175))(12)}{1,470}$$

$$\Rightarrow \gamma \Delta f = 6.73 \text{ ksi}$$

PROBLEM 3.3	TEH	1/4

Strength - I

$P_u = 1.25(1,922) + 1.50(272) + 1.75(744)$

$\Rightarrow P_u = 4,113$ KIPS

$V_{uT} = 1.75(33) + 0.5(23) = 69$ KIPS
$V_{uL} = 1.75(36) + 0.5(30) = 78$ KIPS
$M_{uT} = (M_{uT})_{TOP} + V_{uL} \times 82'$

$= 1.75(432) + 78(82) = 7,152$ ft·k

$M_{uL} = (M_{uL})_{TOP} + V_{uT} \times 82'$

$= 1.75(\overset{396}{432}) + 1.75(3,929)$

$+ 1.25(89) + 69(82)$

$\Rightarrow M_{uL} = 13,338$ ft·k

Strength - III

$P_u = 1.25(1,922) + 1.50(272) = 2,811$ KIPS
$V_{uT} = 1.0(89) + 0.5(23) = 101$ KIPS
$V_{uL} = 1.0(22) + 0.5(30) = 37$ KIPS

$M_{uT} = (M_{uT})_{TOP} + V_{uL} \times 82'$

$= 0 + 37 \times 82 = 3,034$ ft·k

$M_{uL} = (M_{uL})_{TOP} + V_{uT} \times 82'$

$= 1.25(89) + 1.0(801) + 101(82)$

$\Rightarrow M_{uL} = 9,194$ ft·k

Also consider $P_u = 0.9(1,922) + 0(272)$

$= 1,730$ KIPS

PROBLEM 3.3	TEH	2/4

Strength - V   (Wind Speed = 80 mph)

$P_{uL} = 1.25(1,922) + 1.50(272) + 1.35(744)$

$\Rightarrow \quad P_{uL} = 3,815 \text{ KIPS}$

$V_{uLT} = 1.35(33) + 1.0(89)(80/115)^2$
$\qquad + 1.0(18) + 0.5(23)$
$\qquad \Rightarrow V_{uLT} = 118 \text{ KIPS}$

$V_{uLL} = 1.35(36) + 1.0(22)(80/115)^2$
$\qquad + 1.0(8) + 0.5(30)$
$\qquad \Rightarrow V_{uLL} = 82 \text{ KIPS}$

$M_{uLT} = (M_{uT})_{TOP} + V_{uLL} \times 82'$
$\qquad = 1.35(432) + 82(82)$
$\qquad \Rightarrow M_{uT} = 7,307 \text{ ft·k}$

$M_{uLL} = (M_{uLL})_{TOP} + V_{uLT} \times 82'$
$\qquad = 1.35(3,929) + 1.35(396)$
$\qquad + 1.25(89) + 1.0(801)(80/115)^2$
$\qquad + 1.0(324) + 118(82)$

$\qquad \Rightarrow M_{uL} = 16,338 \text{ ft·k}$

PROBLEM 3.3	TEH	3/4

Service - I   (Wind speed = 70 mph)

$P_u = 1.0 (1,922 + 272 + 744) = 2,938^k$

$V_{uT} = 1.0 (33) + 1.0 (89)(70/115)^2$
$\qquad + 1.0 (18) + 1.0 (23)$
$\qquad\qquad \Rightarrow V_{uT} = 107 \text{ kips}$

$V_{uL} = 1.0 (36) + 1.0 (22)(70/115)^2$
$\qquad + 1.0 (8) + 1.0 (30)$
$\qquad\qquad \Rightarrow V_{uL} = 82 \text{ kips}$

$M_{uT} = (M_{uT})_{TOP} + V_{uL} \times 82'$
$\qquad = 1.0 (432) + 82 (82)$
$\qquad\qquad \Rightarrow M_{uT} = 7,156 \text{ ft·k}$

$M_{uL} = (M_{uL})_{TOP} + V_{uT} \times 82'$
$\qquad = 1.0 (89) + 1.0 (3,929) + 1.0 (396)$
$\qquad + 1.0 (801)(70/115)^2 + 1.0 (324)$
$\qquad + 107 (82)$
$\qquad\qquad \Rightarrow M_{uL} = 13,809 \text{ ft·k}$

| PROBLEM 3.3 | | TEH | | | 4/4 |

	P_u	V_{uT}	V_{uL}	M_{uT}	M_{uL}
Str-I	4,113	69	78	7,152	13,338
Str-III	2,811	101	37	3,034	9,194
Str-II	1,730	101	37	3,034	9,194
Str-V	3,815	118	82	7,307	16,338
Ser-I	2,938	107	82	7,156	13,809

Notes: a) P_u, V_{uT}, V_{uL} : KIPS

b) M_{uT}, M_{uL} : FT-KIPS

c) Add column + cap self-
   weight with appropriate
   "DC" load factor to get
   total P_u at column base
   for each limit state

d) For circular columns, compute
   $M_{uT} = (M_{uT}^2 + M_{uL}^2)^{1/2}$

PROBLEM 3.4	TEH	1/4

Use Response 2000 to find:

a) $P_u = 0 \Rightarrow M_n = 2,575$ ft·k

The extreme tension steel is 3.8" from the bottom of the beam. ε_t is the strain in the steel at 3.8" from the bottom = 20.2" below mid-height of the cross-section.

From R2000 strain data, $\varepsilon_t = 0.00863$

$$> \varepsilon_{tl} = 0.005$$

$$\Rightarrow \text{tension-control}, \quad \phi = 0.90$$

$$\phi M_n = 0.90 \times 2,575 = \underline{2,317 \text{ ft·k}}$$

b) $P_u = -2,230$ KIPS $\Rightarrow M_n = 5,225$ FT·K

$$\varepsilon_t = 0.00377$$

$$\varepsilon_{cl} = 0.002 \quad \varepsilon_{tl} = 0.005$$

$\Rightarrow$ Neither tension-control nor compression-control

$$\phi = 0.75 + 0.15 \frac{(.00377 - .002)}{(.005 - .002)}$$

$$\Rightarrow \phi = 0.838$$

$$\phi M_n = 0.838 \times 5,225 = \underline{4,381 \text{ ft·k}}$$

Note: • changed f_u from 90 to 60 for rebar
• Used "parabolic" concrete $\sigma - \varepsilon$
• Strain ε_t obtained from viewing data in Response 2000 (plots give ε at bottom of section, not at rebar level)

PROBLEM 3.4	TEH	2/4

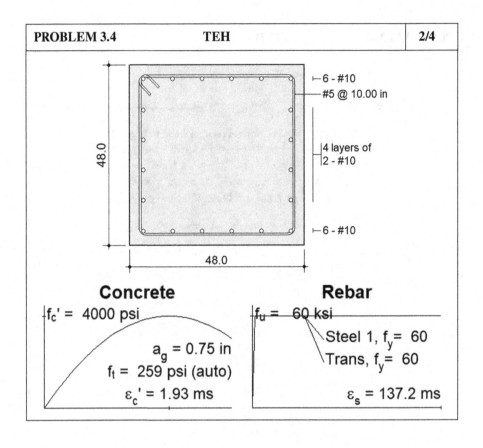

PROBLEM 3.4	TEH	3/4

$P_u = 0$ kips:

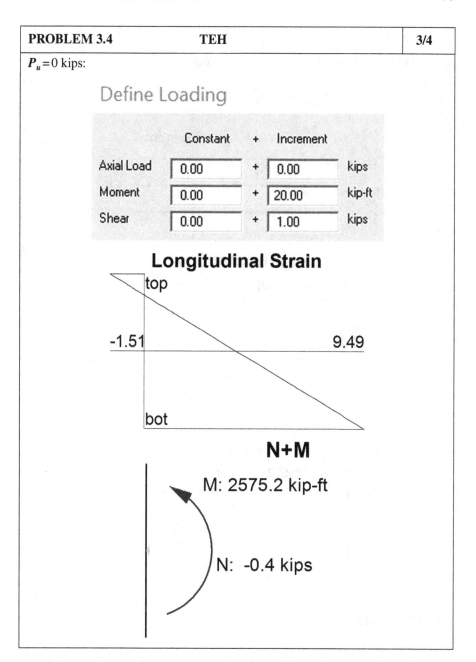

Define Loading

	Constant	+	Increment	
Axial Load	0.00	+	0.00	kips
Moment	0.00	+	20.00	kip-ft
Shear	0.00	+	1.00	kips

Longitudinal Strain

top

-1.51   9.49

bot

N+M

M: 2575.2 kip-ft

N: -0.4 kips

PROBLEM 3.4	TEH	4/4

$P_u = 2,230$ kips compression ("-"in Response 2000)

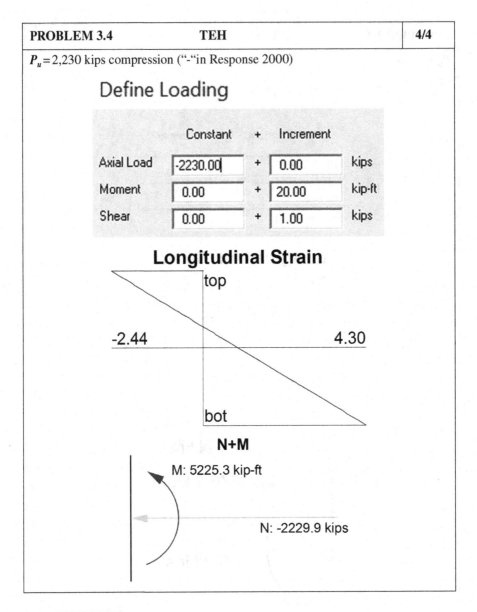

Define Loading

	Constant	+	Increment	
Axial Load	-2230.00	+	0.00	kips
Moment	0.00	+	20.00	kip-ft
Shear	0.00	+	1.00	kips

Longitudinal Strain

top

-2.44   4.30

bot

N+M

M: 5225.3 kip-ft

N: -2229.9 kips

3.2 EXERCISES

Exercise 3.1.

Figure E3.1 shows a section of bridge deck overhang with loads from various sources. Determine the Service I, Strength I, and Extreme Event II limit state tension (T) and moment (M) in the bridge deck. The parapet is 1 ft wide over the entire parapet height.

- $P_{DC} = 0.400$ klf from the parapet
- $P_{LL} = 3.5$ klf from live load (impact not included)

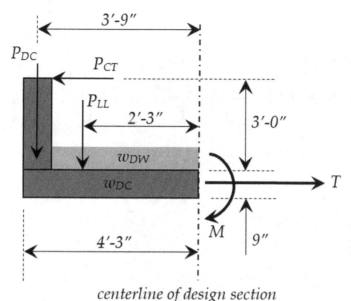

centerline of design section

FIGURE E3.1 Exercise E3.1.

- $P_{CT} = 6.2$ klf from vehicular collision
- $w_{DW} = 35$ psf from future overlay allowance
- $w_{DC} = $ deck weight of 9-inch-thick concrete (150 pcf)

Exercise 3.2.

Figure E3.2 depicts the loading on an intermediate pier for a continuous span bridge. The loads are given as follows:

Vertical reactions:
- $R_{DC} = 407$ kips per girder (same for interior and exterior girders)
- $R_{DW} = 79$ kips per girder (same for interior and exterior girders)
- $R_{LL} = 160$ kips per lane from uniform live load
- $R_{LL} = 115$ kips per lane from truck live load

Wind loads:
- $V_{IWS} = 63$ kips for Strength III limit state
- $V_{IWS} = 30$ kips for Strength V limit state
- $V_{IWS} = 23$ kips for Service I limit state
- $V_{IWS} = 35$ kips for Service IV limit state
- $V_{2WS} = 6$ kips for Strength III and Strength V limit states
- $P_{WS} = 144$ kips for Strength III limit state
- $P_{WS} = 72$ kips for Service IV limit state

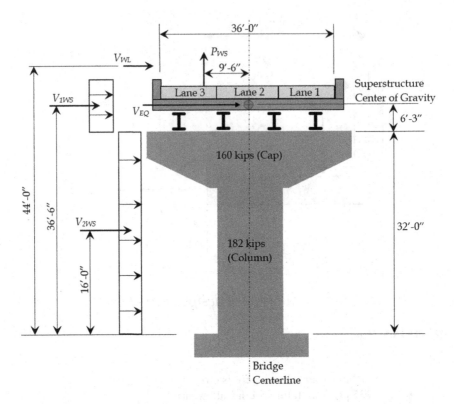

FIGURE E3.2 Exercise E3.2.

Seismic load: $V_{EQ} = 496$ kips

Determine the required shear, axial force, and moment at the base of the pier column for each of the following limit states:

- Strength I
- Strength III
- Strength V
- Service I
- Service IV
- Extreme Event I (only one lane of live load and $\gamma_{EQ} = 1.0$)

4 Deck and Parapet Design

Traditional deck design is accomplished assuming the deck to be a continuous beam supported at the girders, with primary reinforcement placed transverse to the direction of traffic. This requires establishing an effective strip width of deck for resisting live load. The equivalent strip method of deck design is covered in detail in Section 4.6.2 of the 9th edition of the AASHTO LRFD BDS. For overhang deck design, the wheel load from the design truck is to be placed 1 foot from the face of the parapet. For all other considerations (for example, live load distribution), the wheel load is placed 2 feet from the parapet face.

Section 9.7.3.2 of the AASHTO LRFD BDS requires distribution reinforcement in the bottom of girder-bridge decks in the longitudinal direction, with main deck reinforcement transverse to the direction of traffic. Such distribution reinforcement is to be determined as a percentage, p, of the primary (bottom transverse) reinforcement, as given by Equation 4.1, a function of the effective span, S. For prestressed and steel girder bridges, S may conservatively be taken to be equal to the girder spacing.

$$p = \frac{220}{\sqrt{S}} \leq 67 \qquad (4.1)$$

Equivalent strip widths are summarized in Table 4.1 for cast-in-place concrete decks on steel or concrete girders. The girder spacing, S, is used to determine interior strip widths. The distance, X, from the load to the point of support is used to determine strip width in the deck overhang. Note that both X and S must be expressed in feet, while the resulting strip width is in inches. As an alternative design method for deck overhangs at the Strength limit state, Section 3.6.1.3.4 of the AASHTO LRFD BDS permits the outer wheel load to be replaced by a uniformly distributed load of 1.0 kip per foot located 1 foot from the face of the barrier, provided (a) the barrier is continuous and (b) the distance from the exterior girder centerline to the face of the barrier, d_e, does not exceed 6 feet.

4.1 PARAPET DESIGN

Parapet design for bridges includes accommodating vehicular crash loads (CT), both vertical and horizontal. Various test-level parameters used for crash load design of parapets are summarized in Table 4.2. Horizontal forces, F_t and F_L, are applied at the top of the parapet.

Parapet design includes minimum height requirements, as shown in Table 4.2, as well as resistance requirements. Parapet resistance, R_w, given in Equation 4.4 for an interior segment of parapet, or in Equation 4.5 for an end segment, must equal or exceed F_t.

TABLE 4.1
Equivalent Strip Width

Condition	Equivalent Strip Width, inches
Overhang, top of deck (negative moment)	$45 + 10.0X$
Interior, bottom of deck (positive moment)	$26 + 6.6S$
Interior, top of deck (negative moment)	$48 + 3.0S$

TABLE 4.2
Crash Load (CT) Requirements

Parameter	TL-1	TL-2	TL-3	TL-4	TL-5	TL-6
F_t, transverse (kips)	13.5	27.0	54.0	54.0	124.0	175.0
F_L, longitudinal (kips)	4.5	9.0	18.0	18.0	41.0	58.0
F_v, vertical (kips)	4.5	4.5	4.5	18.0	80.0	80.0
L_t and L_L (feet)	4.0	4.0	4.0	3.5	8.0	8.0
L_v (feet)	18.0	18.0	18.0	18.0	40.0	40.0
H_e (minimum, inches)	18.0	20.0	24.0	32.0	42.0	56.0
Minimum rail height, H (ft)	27.0	27.0	27.0	32.0	42.0	90.0

L_c is a critical yield line length, in feet, determined using either Equation 4.2 for an interior segment of parapet, or Equation 4.3 for an end segment of parapet. H is the actual height of the parapet, in feet. The moment, M_b, is the flexural resistance of any beam on top of the parapet, has units of ft-kips, and is often zero. M_w is the flexural resistance of the parapet about a vertical axis, in units of ft-kips.

The moment, M_C, is equal to the flexural resistance of the parapet base about an axis in the direction of traffic, in units of ft-kips per foot. For parapets with excess capacity, this provision certainly penalizes the design of the deck overhang reinforcement, as will be evident when design cases for the deck overhang are discussed.

$$L_c = \frac{L_t}{2} + \sqrt{\left(\frac{L_t}{2}\right)^2 + \frac{8H(M_b + M_w)}{M_c}}, \text{ interior segment} \qquad (4.2)$$

$$L_c = \frac{L_t}{2} + \sqrt{\left(\frac{L_t}{2}\right)^2 + \frac{H(M_b + M_w)}{M_c}}, \text{ end segment} \qquad (4.3)$$

$$R_w = \left(\frac{2}{2L_c - L_t}\right)\left(8M_b + 8M_w + \frac{M_c L_c^2}{H}\right), \text{ interior segment} \qquad (4.4)$$

$$R_w = \left(\frac{2}{2L_c - L_t}\right)\left(M_b + M_w + \frac{M_c L_c^2}{H}\right), \text{end segment} \qquad (4.5)$$

4.2 DECK OVERHANG DESIGN

Section A13.4 of Appendix A13 in the AASHTO-LRFD-BDS requires that deck overhangs be designed for three cases.

A. Extreme Event II limit state: horizontal crash forces (CT) with load factor $\gamma_{CT}=1.00$, live load plus impact forces (LLL+IM) with load factor $\gamma_{LL}=0.50$, and dead loads (DC, DW) with a load factor $\gamma_{DC}=\gamma_{DW}=1.00$.
B. Extreme Event II limit state: vertical crash forces (CT) with load factor $\gamma_{CT}=1.00$, live load plus impact forces (LLL+IM) with load factor $\gamma_{LL}=0.50$, and dead loads (DC, DW) with a load factor $\gamma_{DC}=\gamma_{DW}=1.00$.
C. Strength I limit state: live load plus impact (LL+IM) with load factor $\gamma_{LL}=1.75$, dead loads (DC) with load factor $\gamma_{DC}=1.25$, dead load (DW) with load factor $\gamma_{DW}=1.50$.

In bridges with concrete parapets, for design case A, the overhang may be designed for a tensile force, T, acting simultaneously with the moment, M_C. The tension, T, is given by Equation 4.6 and is expressed in units of kips per foot.

$$T = \frac{R_w}{L_c + 2H} \qquad (4.6)$$

This approach, although recommended in Section 13 of the AASHTO LRFD-BDS, is extremely conservative in that contributions of slab resistance perpendicular to traffic on the overhang are completely ignored, but likely to be significant in reality. Current research (National Cooperative Highway Research Program – NCHRP Project 12–119) is aimed at more accurately predicting the behavior of deck overhangs.

4.3 INTERIOR BAY DECK DESIGN

The design of reinforcement in the deck over interior girders (negative moment, top mat of steel) and mid-span between girders (positive moment, bottom mat of steel) is fairly straightforward, compared to overhang design. Appendix A4 of the AASHTO LRFD BDS provides tables for positive and negative live load moment in bridge decks, expressed in units of ft-kips per foot. The deck design table already has multi-presence and impact factors incorporated in the tabulated moments. Simplified design using the table may be used as long as the following conditions are satisfied:

- the deck is supported by three or more girders;
- the distance between centerlines of exterior girders is no less than 14 feet;

- the girders are parallel; and
- the overhang must be designed separately as outlined in Section 4.2 above.

If the tables are not available, approximate regression equations may be useful in the following form shown in Equation 4.7 for cases in which the girder spacing is no less than 7.5 feet and no greater than 15 feet. Coefficients for the equation are summarized in Table 4.3.

$$M = a + b \times S + c \times S^2 \tag{4.7}$$

Live load moments from the table may be combined, using appropriate load factors, with dead load moments computed using simplified approximations, such as those presented in Equations 4.8 and 4.9. S, again, is the girder spacing. The design section distance, d_s, is the distance from the centerline of the girder to the design section for negative moment and is taken to be equal to one-fourth of the flange width for steel I-girders, and one-third of the flange width for concrete I-girders.

$$M^+_{DC,DW} = \frac{w_{DC,DW}S^2}{11} \tag{4.8}$$

$$M^-_{DC,DW} = \frac{w_{DC,DW}\left(S - 2d_s\right)^2}{9} \tag{4.9}$$

TABLE 4.3
Coefficients for Slab Moment Regression Equation

	7.5 feet $\leq S \leq$ 9.25 feet			9.25 feet $< S \leq$ 15 feet		
	a	b	c	a	b	c
Positive M	3.86548	−0.09095	0.04000	−0.56975	0.89761	−0.01519
Negative M (d_s=0)	−1.35893	1.55905	−0.07238	−10.87797	2.42897	−0.05550
Negative M (d_s=3 inches)	−1.89060	1.49048	−0.06857	−11.08878	2.31935	−0.05105
Negative M (d_s=6 inches)	−1.65375	1.24214	−0.05429	−11.32318	2.21442	−0.04681
Negative M (d_s=9 inches)	0.04821	0.62714	−0.01714	−11.18834	2.05032	−0.04023
Negative M (d_s=12 inches)	−7.45810	2.27333	−0.11429	−8.92069	1.55216	−0.02072
Negative M (d_s=18 inches)	−21.02292	4.80500	−0.23333	−3.14296	0.68492	0.00368
Negative M (d_s=24 inches)	−6.07560	1.19667	−0.02095	−6.27981	1.24680	−0.02410

In cases where the cross-sectional geometry lies outside the limits for simplified design, or for cases where more precise deck moments are needed, a continuous beam model (supports at the girder centerlines) with overhangs may be developed, with the moving load consisting of side-by-side vehicles comprised of wheel loads from the HL-93 truck. The magnitude of such wheel loads would need to be 16 kips (32-kip axle = 16-kip wheel load) × 1.33 (impact) × 1.2 (multi-presence) = 25.5 kips. Uniform loads from the HL-93 lane load, the deck self-weight, and the overlay allowance must be included in such models as well. Resulting moments from the moving load analysis clearly need to be divided by the effective strip width prior to combining with dead load effects.

Design flexural resistance, ϕM_n, must be greater than the Strength limit state-required flexural resistance, M_u. Refer to Chapters 2 and 3 for previous discussions on appropriate load (γ) factors and resistance (ϕ) factors.

To control deck cracking, Section 5.6 of the AASHTO LRFD BDS requires that bar spacing in the extreme tension layer of reinforcement does not exceed s as given by Equation 4.10.

$$s \le \frac{700\gamma_e}{\beta_s f_{ss}} - 2d_c \tag{4.10}$$

$$\beta_s = 1 + \frac{d_c}{0.7(h - d_c)} \tag{4.11}$$

β_s is the ratio of strain at the extreme tension face to strain in the centroid of the reinforcement layer closest to the tension face, and d_c is the distance from the extreme tension face to the centroid of the reinforcement closest to the tension face. The overall height of the member is h. The Service limit state stress in the reinforcement is f_{ss}, which must not exceed $0.60 \times f_y$. The exposure factor, γ_e, is 1.00 for Class 1 exposure conditions and 0.75 for Class 2 exposure conditions. Class 1 exposure corresponds to an estimated crack width of 0.017 inches. Class 2 exposure corresponds to an estimated crack width of 0.013 inches, as the exposure factor is directly proportional to the estimated crack width. Decks are one example of a case where it may be advisable to use Class 2 exposure for crack control, given the susceptibility to corrosion for typical deck top conditions.

For Strength limit state flexural considerations, bridge decks are typically tension-controlled elements with a resistance factor, $\phi = 0.90$. Nevertheless, such assumptions should always be verified and any adjustments to resistance incorporated.

The AASHTO LRFD BDS clear cover requirements are found in Section 5.10.1. For bridge decks with uncoated reinforcing steel subject to tire stud or chain wear, the minimum clear cover is 2.50 inches. For such a situation, but with epoxy-coated reinforcement, the minimum clear cover is 2.00 inches. For the bottom of cast-in-place slabs, the minimum clear cover is 1.00 inch for No. 11 bars and smaller. The maximum bar size in typical bridge decks is #6 for the outer mats perpendicular to traffic. For precast, prestressed girders made continuous and requiring large amounts of longitudinal steel in the deck to resist negative moments at interior supports, longitudinal bar sizes as large as #9 may be required. Congested reinforcement geometries should be avoided and the minimum spacing between the top and bottom longitudinal mats of reinforcing must be no less than either (a) 1.00 inch and (b) the longitudinal bar diameter. For an 8.25-inch- thick bridge deck with #6 transverse bars top and bottom, and with #6 bottom longitudinal bars, the maximum size of the top longitudinal bar which may be used while maintaining the 1-inch clear distance, is a #8 bar, assuming 2.50-inches top clear cover and 1.00-inch bottom clear cover.

Transverse mats of deck reinforcing are typically placed outside the longitudinal mats to provide advantageous effective depth for flexural design of the transverse reinforcement. Consideration of compression reinforcement for design of the transverse deck reinforcement, although often ignored, may prove beneficial for analysis at the Service and Strength limit states.

4.4 SOLVED PROBLEMS

Problem 4.1

For the bridge cross-section shown in Figure P4.1, design the transverse deck reinforcement and the bottom mat of longitudinal distribution reinforcement. Use $f'_c = 4$ ksi, $f_y = 60$ ksi, and Exposure 2 for the top mat of transverse reinforcement. The girders are steel, welded-plate girders with top flange width, $b_f = 18$ inches. Also assess the adequacy of the parapet for TL-4 criteria. For Extreme Event considerations, parapet resistance values have been determined to be:

- $M_W = 20.95$ ft-k
- $M_C = 16.82$ ft-k/ft

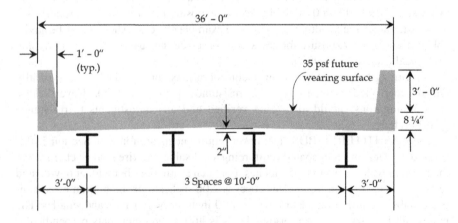

FIGURE P4.1   Problem P4.1.

The parapets weigh 0.38 klf each with a center of gravity located 7.5 inches from the deck edge. Use 2.5-inches clear cover for top bars and 1-inch clear cover for bottom bars.

Problem 4.2

Determine M_W and M_C for the parapet shown in Figure P4.2. Assess the adequacy of the parapet for various TL criteria. Use $T_w = 6$ inches, $f'_c = 5$ ksi, and $f_y = 60$ ksi.

Problem 4.3

A 9-inch-thick bridge deck with 2.50 inches top clear cover and 1.00 inches bottom clear cover has #5 transverse bars spaced at 7 inches on centers, top and bottom. Concrete strength, f'_c equals 4 ksi. Yield stress for the epoxy-coated reinforcement is 60 ksi. Determine each of the following for both positive flexure (tension in the bottom of the deck) and negative flexure (tension in the top of the deck).

- ϕM_n, design flexural resistance at the Strength limit state
- I_{cr}, the cracked section moment of inertia
- f_{ss} as a function of Service limit state moment, $M_{u\text{-}SER}$
- β_s

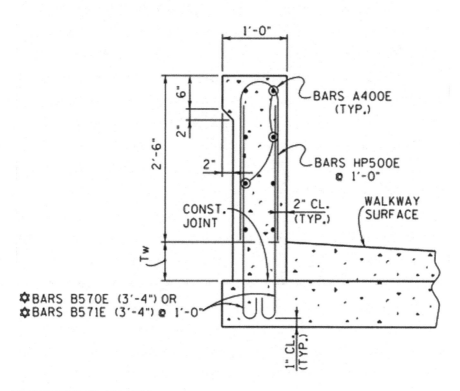

FIGURE P4.2   Problem P4.2.

PROBLEM 4.1	TEH	1/11

Steel Girder, design section.

$$d_s = b_f/4 = 18/4 = 4.5''$$

Deck weight $= (8.25/12)(0.150)$

$$= 0.103 \text{ KLF/FT}$$

Overlay $= 0.035 \text{ KLF/FT}$

<u>Interior M^+</u>

$$M^+_{DC} = 0.103(10)^2/11 = 0.94 \text{ ft·k/ft}$$

$$M^+_{DW} = 0.035(10)^2/11 = 0.32 \text{ ft·k/ft}$$

$$M^+_{LL+IM} = 6.89 \text{ ft·k/f (table)}$$

Service I :

$$M^+_{SER-I} = 0.94 + 0.32 + 6.89$$

$$= \underline{8.15 \text{ ft·k/ft}}$$

Strength I :

$$M^+_{STR-I} = 1.25(0.94) + 1.50(0.32)$$
$$+ 1.75(6.89)$$

$$= \underline{13.71 \text{ ft·k/ft}}$$

PROBLEM 4.1	TEH	2/11

Interior M^-

$$M^-_{DC} = 0.103 \left(10 - \frac{9}{12}\right)^2 / 9$$

$$= 0.98 \text{ ft} \cdot \text{k} / \text{ft}$$

$$M^-_{DW} = 0.035 \left(10 - \frac{9}{12}\right)^2 / 9$$

$$= 0.33 \text{ ft} \cdot \text{k} / \text{ft}$$

$$M^-_{LL+IM} = \frac{1}{2}(6.99 + 6.13)$$

$$= 6.56 \text{ ft} \cdot \text{k} / \text{ft}$$

Service I :

$$M^-_{SER-I} = 0.98 + 0.33 + 6.56$$

$$= \underline{7.87 \text{ ft} \cdot \text{k} / \text{ft}}$$

Strength - I :

$$M^-_{STR-I} = 1.25 (0.98)$$
$$+ 1.50 (0.33)$$
$$+ 1.75 (6.56)$$

$$= \underline{13.20 \text{ ft} \cdot \text{k} / \text{ft}}$$

PROBLEM 4.1	TEH	3/11

Overhang M^-

$P = 16^K \times 1.33 \times 1.2$
$\qquad (IM) \quad (m)$

$d_s = 4.5''$

$\Rightarrow P = 25.5^K$

$X = 3' - 2' - 4.5/12$

$\Rightarrow X = 0.625'$

$M_{DC}^- = 0.103 (2.625)^2 / 2$
$\qquad + 0.38 (2.625 - 7.5/12)$

$\qquad \Rightarrow M_{DC}^- = 1.12 \ KFT/FT$

$M_{DW}^- = 0.035 (2.625 - 1)^2 / 2$

$\qquad \Rightarrow M_{DW}^- = 0.05 \ ft \cdot K/ft$

Strip width $= 45 + 10(0.625)$
$\qquad = 51.25'' = 4.27'$

$M_{LL+IM}^- = 25.5^K (0.625') / 4.27$

$\qquad = 3.73 \ ft \cdot K/ft$

PROBLEM 4.1	TEH	4/11

Transverse $\angle T$.

$$TL-4 \rightarrow \bar{F_t} = 54 \text{ kips}$$
$$L_t = 3.5 \text{ ft}$$

End Segment of Parapet:

$$L_c = \frac{3.5}{2} + \sqrt{\left(\frac{3.5}{2}\right)^2 + \frac{3(20.95)}{16.82}} = 4.36'$$

$$R_w = \frac{2}{2(4.36)-3.5}\left(20.95 + \frac{16.82 \times 4.36^2}{3}\right)$$

$$\Rightarrow R_w = 48.9 \text{ kips} < F_t$$

$\rightarrow$ End Segment requires strengthening.

Interior Segment of Parapet:

$$L_c = \frac{3.5}{2} + \sqrt{\left(\frac{3.5}{2}\right)^2 + \frac{24(20.95)}{16.82}} = 7.49'$$

$$R_w = \frac{2}{2(7.49)-3.5}\left(8 \times 20.95 + \frac{16.82 \times 7.49^2}{3}\right)$$

$$\Rightarrow R_w = 84.0 \text{ k} > F_t$$

$\rightarrow$ Interior parapet segment OK

PROBLEM 4.1	TEH	5/11

Overhang Design Case 1.

$$\overline{M_u} = M_c = 16.82 \text{ ft-k} / \text{ft}$$

$$T = \frac{84.0}{7.49 + 2(3)} = 6.23 \text{ k} / \text{ft}$$

Overhang Design Case 2.

$F_r = 18$ kips

$L_v = 18$ ft

Extreme Event II

$$M_{CT}^- = 18^k / 18 \text{ ft} (2') = 2 \text{ ft·k/ft}$$

(assume vertical CT load is applied at top of parapet on the inside edge)

$$M_{EXTR}^- = 1.0(1.12 + 0.05) + 0.5(3.73) + 2.00$$

$$= 5.04 \text{ ft·k} / \text{ft}$$

Overhang Design Case 3.

$$M_{SER-I}^- = 1.12 + 0.05 + 3.73 = 4.90 \frac{\text{ft·k}}{\text{ft}}$$

$$M_{STR-I}^- = 1.25(1.12) + 1.5(0.05) + 1.75(3.73)$$

$$= 8.00 \text{ ft·k} / \text{ft}$$

PROBLEM 4.1	TEH	6/11

After several iterations try

  #5 bars spaced 5" on center, top
  #5 bars spaced 8" on center, bott

Response 2000 file "CEE4380-Problem 4-1"

Extreme Event $M^- = 16.82$ ft·k/ft

$T = 6.23$ k/ft

$\Rightarrow \phi M_n = 1.0 \times 16.9$

$= 16.9$ ft·k/ft

> 16.82, ok

Strength $M^- = 13.2$ ft·k/ft

$\Rightarrow \phi M_n = 0.9 \times 18.4$

$= 16.6$ ft·k/ft

> 13.2, ok

Strength $M^+ = 13.71$ ft·k/ft

$\Rightarrow \phi M_n = .9 \times 21.4$

$= 19.3$ ft·k/ft

> 13.71, ok

PROBLEM 4.1	TEH	7/11

Service $M^- = 7.87$ ft·k/ft

$f_{ss} = 26.1$ ksi

$d_c = 2.5 + 5/16 = 2.81''$

$B_s = 1 + \dfrac{2.81}{.7(8.25 - 2.81)}$

$= 1.738$

$S \le \dfrac{700(.75)}{1.738(26.1)} - 2(2.81)$

$S \le 5.95''$, $S_{actual} = 5''$, ok

Service $M^+ = 8.15$ ft·k/ft

$f_{ss} = 30.2$ ksi

$d_c = 1 + 5/16 = 1.31''$

$B_s = 1 + \dfrac{1.31}{.7(8.25 - 1.31)}$

$= 1.270$

$S \le \dfrac{700(1.0)}{1.27(30.2)} - 2(1.31)$

$S \le 15.6''$, $S_{actual} = 8''$, ok

PROBLEM 4.1	TEH	8/11

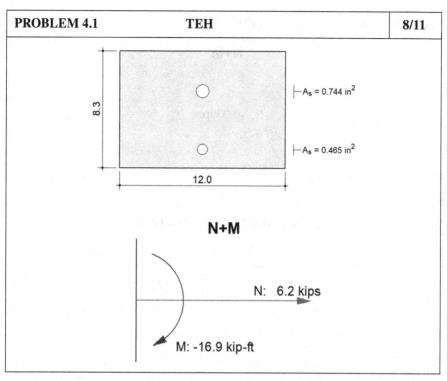

$A_s = 0.744 \text{ in}^2$

$A_s = 0.465 \text{ in}^2$

8.3

12.0

N+M

N: 6.2 kips

M: -16.9 kip-ft

PROBLEM 4.1	TEH	9/11

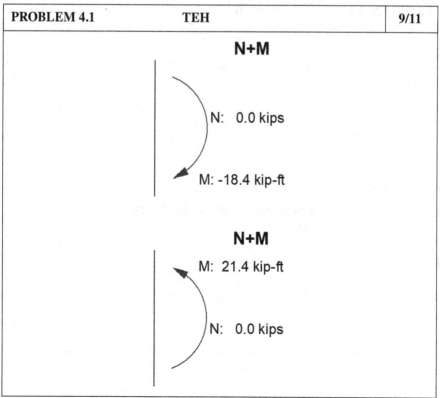

N+M

N: 0.0 kips

M: -18.4 kip-ft

N+M

M: 21.4 kip-ft

N: 0.0 kips

PROBLEM 4.1	TEH	10/11

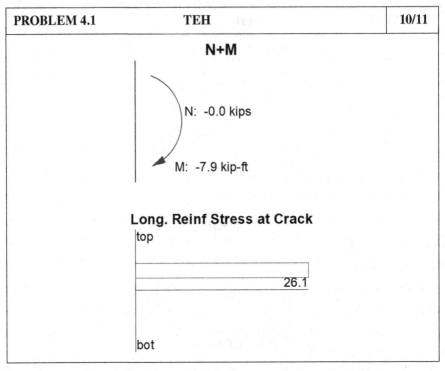

PROBLEM 4.1	TEH	11/11

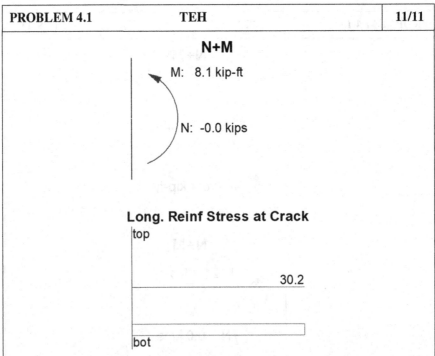

Distribution steel (bottom mat, longitudinal):

$$p = \frac{220}{\sqrt{10}} = 69.6 > 67 \rightarrow \text{use 67\% of No. 5 @ 8''}$$

Use No. 5 at 12 inches on center

PROBLEM 4.2	TEH	1/3

Response 2000 file, "CEE4380-Problem-4.2-MW.rsp"

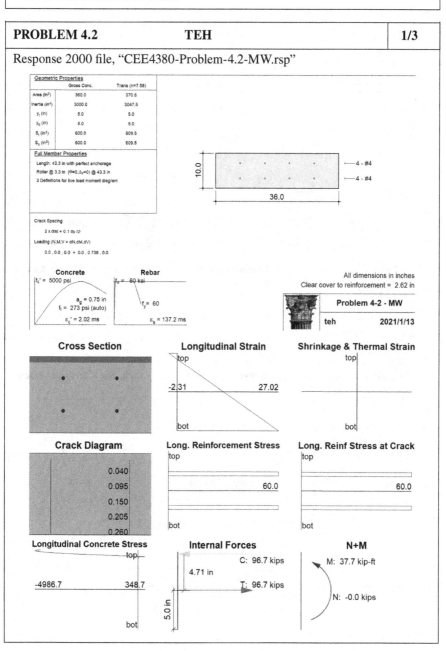

PROBLEM 4.2   **TEH**   **2/3**

Response 2000 file, "CEE4380-Problem-4.2-MC.rsp"

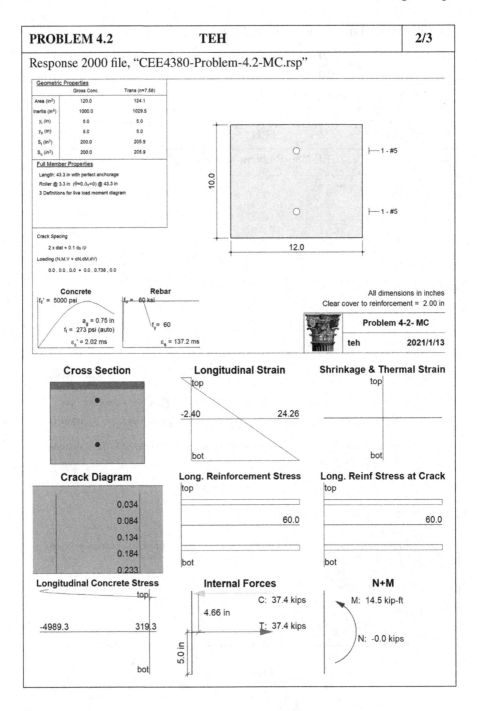

Geometric Properties

	Gross Conc.	Trans (n=7.58)
Area (in²)	120.0	124.1
Inertia (in⁴)	1000.0	1029.5
y_t (in)	5.0	5.0
y_b (in)	5.0	5.0
S_t (in³)	200.0	205.9
S_b (in³)	200.0	205.9

Full Member Properties

Length: 43.3 in with perfect anchorage
Roller @ 3.3 in (θ=0,Δ_y=0) @ 43.3 in
3 Definitions for live load moment diagram

Crack Spacing

2 x dist + 0.1 d_b /ρ

Loading (N,M,V + dN,dM,dV)

0.0 , 0.0 , 0.0 + 0.0 , 0.738 , 0.0

Concrete
f_c' = 5000 psi
a_g = 0.75 in
f_t = 273 psi (auto)
ε_c' = 2.02 ms

Rebar
f_u = 60 ksi
f_y = 60
ε_s = 137.2 ms

All dimensions in inches
Clear cover to reinforcement = 2.00 in

Problem 4-2- MC
teh   2021/1/13

10.0
12.0
1 - #5
1 - #5

Cross Section

Longitudinal Strain
top
-2.40   24.26
bot

Shrinkage & Thermal Strain
top
bot

Crack Diagram
0.034
0.084
0.134
0.184
0.233

Long. Reinforcement Stress
top
60.0
bot

Long. Reinf Stress at Crack
top
60.0
bot

Longitudinal Concrete Stress
top
-4989.3   319.3
bot

Internal Forces
C: 37.4 kips
4.66 in
T: 37.4 kips
5.0 in

N+M
M: 14.5 kip-ft
N: -0.0 kips

PROBLEM 4.2	TEH	3/3

From the Response 2000 analyses:

$$M_W = 37.7 \text{ ft-kips}$$

$$M_C = 14.5 \text{ ft-k/ft}$$

Consider TL-3 criteria (highest level for which H is at least equal to H_{min}):

$$F_t = 54 \text{ kips}$$

$$L_t = 4.0 \text{ feet}$$

$H = 27$ inches (min), $H = 30$ inches (actual), OK

For an interior parapet segment:

$$L_c = \frac{4}{2} + \sqrt{\left(\frac{4}{2}\right)^2 + \frac{8(2.5)(0+37.7)}{14.5}} = 9.48 \text{ ft}$$

$$R_w = \left(\frac{2}{2(9.48)-4}\right)\left(8(37.7) + \frac{14.5(9.48)^2}{2.5}\right) = 110 \text{ kips}$$

For an end segment of the parapet:

$$L_c = \frac{4}{2} + \sqrt{\left(\frac{4}{2}\right)^2 + \frac{(2.5)(0+37.7)}{14.5}} = 5.24 \text{ ft}$$

$$R_w = \left(\frac{2}{2(5.24)-4}\right)\left(37.7 + \frac{14.5(5.24)^2}{2.5}\right) = 61 \text{ kips}$$

For both interior and end parapet segments, $R_w > F_t \to$ OK.

TL-3 criteria are satisfied for the parapet geometry and reinforcement. TL-4 and beyond require parapet heights greater than that provided and are thus not satisfied, regardless of any computed R_w value.

PROBLEM 4.3	TEH	1/4

$$A_s = A'_s = 0.31 \left(12/7\right) = 0.531 \; in^2/ft$$

$$n = \frac{29,000}{1,820\sqrt{4}} = 7.97, \; say \; n = 8$$

Positive Flexure (bottom tension)

$$A_s f_y = 0.531 \times 60 = 31.86 \; kips$$

$$.85(4)(12)(2.81) = 114.6 \; kips$$

$$\Rightarrow \text{"compression" (top) steel}$$
$$\text{is actually in tension}$$

Assume all steel yields $\dot{\imath}$ verify:

$$a = 0.85 \, c$$

$$.85(4)(a)(12) = 0.531(60)(2)$$

$$\Rightarrow a = 1.56''$$

$$\Rightarrow c = 1.84''$$

$$\varepsilon'_s = \frac{0.003}{1.84} (2.81 - 1.84) = 0.0016$$

$$< \varepsilon_y = 0.00207, \; \text{top steel} \atop \text{does not} \atop \text{yield}$$

PROBLEM 4.3	TEH	2/4

$$0.85(4)(a)(12) = \frac{.003}{c}\left(2.81-c\right)\left(29,000\right)\left(0.531\right)$$
$$+ 60\left(0.531\right)$$

$$34.68C = \frac{129.81}{C} - 46.2 + 31.86$$

$$\Rightarrow C = 1.74'' \qquad \mathcal{E}_s' = 0.00184$$
$$a = 1.48'' \qquad \mathcal{E}_s = 0.0103 > \mathcal{E}_y$$

$$\overline{F_s'} = 0.00184 \times 29,000 \times 0.531$$
$$= 28.3 \text{ KIPS}$$

$$\overline{F_s} = 60 \times .531 = 31.9 \text{ KIPS}$$

$$\phi M_n^+ = 0.9 \left[28.3(2.81-1.48/2) + 31.9(7.29-1.48/2)\right]$$
$$= 252 \frac{\text{in} \cdot \text{k}}{\text{ft}} = \underline{21.0 \text{ ft} \cdot \text{k}/\text{ft}}$$

For the service limit state, assume
the top steel is in compression &
verify: $(n-1)A_s' = 7 \times .531 = 3.72 \text{ in}^2$
$\qquad \wedge A_s = 8 \times .531 = 4.25 \text{ in}^2$

$$(12)(kd)(kd/2) + 3.72(kd-2.81) = 4.25(7.69-kd)$$
$$6(kd)^2 + 3.72(kd) - 10.45 = 32.68 - 4.25(kd)$$

$$\Rightarrow kd = 2.10'' < 2.81''$$
$$\Rightarrow \text{top bars are in tension}$$
$$\text{at service limit state}$$
$$\text{as well.}$$

PROBLEM 4.3	TEH	3/4

$$6(kd)^2 = 4.25(2.81 - kd) + 4.25(7.69 - kd)$$

$$\Rightarrow kd = 2.11 \text{ inches}$$

$$I_{cr} = \frac{12(2.11)^3}{3} + 4.25(0.7^2 + 5.58^2)$$

$$\Rightarrow \underline{I_{cr}^+ = 172 \text{ in}^4/ft}$$

$$\beta_s^+ = \frac{9 - 2.81}{7.69 - 2.81} = \underline{1.27}$$

$$f_{ss}^+ = \frac{12 M_{u\text{-}ser}(7.69 - 2.81)(8)}{172}$$

$$\underline{f_{ss}^+ = 2.72 M_{u\text{-}ser}} \quad \left(\begin{array}{c} M_{u\text{-}ser}, ft \cdot k \\ f_{ss}, ksi \end{array}\right)$$

Use Response 2000 for ϕM_n^-, I_{cr}^-, β_s^-, f_{ss}^-

$$\phi M_n^- = 0.9 \times 15.9 = 14.3 \, ft \cdot k/ft$$

$$f_{ss}^- = 3.96 M_{u\text{-}ser}^- \quad \left(\begin{array}{c} M_{u\text{-}ser}, ft \cdot k \\ f_{ss}, ksi \end{array}\right)$$

Response 2000 file "CEE4380-P04.03" is shown on the following page for the negative moment analysis.

PROBLEM 4.3	TEH	4/4

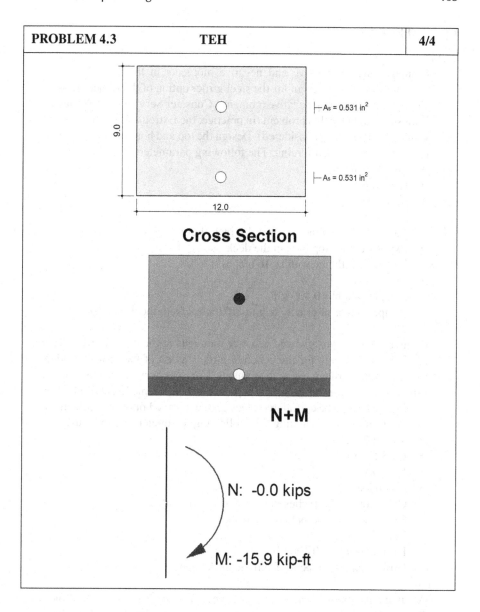

As = 0.531 in²

As = 0.531 in²

9.0

12.0

Cross Section

N+M

N: -0.0 kips

M: -15.9 kip-ft

4.5 EXERCISES

E4.1

Compute interior positive and negative moments in the deck for transverse reinforcement design for the steel girder option of the Project Bridge. Compute the overhang negative moments. Consider Service I and Strength I limit states only for this problem (in practice, the Extreme Event limit states would also have to be considered). Design the top and bottom reinforcement mats of transverse reinforcing. The following parameters are to be used:

- $S = 9$ ft 3 inches
- $C = 3$ ft 4 inches
- $f'_c = 4$ ksi
- $f_y = 60$ ksi
- Clear cover = 2.5 inches (top), 1.0 inch (bottom)
- Exposure 2 for top bars in the deck
- Steel girder flange width = 16 inches
- Use #5 bars
- Parapet weight = 0.400 klf
- Parapet center of gravity (c.g.) is 10 inches from the deck edge

E4.2

Compute interior positive and negative moments in the deck for transverse reinforcement design for the concrete girder option of the Project Bridge. Compute the overhang negative moments. Consider Service I and Strength I limit states only for this problem (in practice, the Extreme Event limit states would also have to be considered). Design the top and bottom reinforcement mats of transverse reinforcing. The following parameters are to be used:

- $S = 9$ ft 3 inches
- $C = 3$ ft 4 inches
- $f'_c = 4$ ksi
- $f_y = 60$ ksi
- Clear cover = 2.5 inches (top), 1.0 inch (bottom)
- Exposure 2 for top bars in the deck
- Use #5 bars
- Parapet weight = 0.400 klf
- Parapet c.g. is 10 inches from the deck edge

E4.3

Assume a parapet with no excess resistance ($R_w = F_t$) is used for the Project Bridge. Compute overhang design moments at the Extreme Event limit state for the steel girder option of the Project Bridge. Assume that the ratio $M_w/M_c = 2.5$, and that $M_b = 0$. Compare the design moment to the Strength limit state moments from Exercise 4.1.

5 Distribution of Live Load

Live load distribution factors permit girder design to be completed using a line girder analysis (i.e., a single, continuous beam model) rather than a 3-dimensional or grillage computer model. This greatly simplifies girder design and the interpretation of results.

AASHTO includes at least three methods for computing live load distribution factors:

1. AASHTO equations
2. The lever rule
3. The rigid cross-section method

The decreasing likelihood of simultaneous lanes being loaded in the most disadvantageous arrangement, as the number of loaded lanes increases, is incorporated into AASHTO live load design through the application of a multi-presence factor, m. The multi-presence factor is applied as follows (with the exception of the Fatigue limit state, for which multi-presence factors are not applied):

- 1 loaded lane, $m = 1.20$
- 2 loaded lanes, $m = 1.00$
- 3 loaded lanes, $m = 0.85$
- 4 or more loaded lanes, $m = 0.65$

Several additional definitions will be necessary prior to presenting the various methods available in the AASHTO BDS for live load distribution in I-girder bridges.

- S = girder spacing, feet
- t_s = concrete deck thickness, inches
- K_g = longitudinal stiffness parameter, in^4
- L = span length, feet
- e = correction factor
- d_e = distance from centerline of exterior I-girder to face of parapet, feet
- E_B = Young's modulus for the girder concrete, ksi
- E_D = Young's modulus for the deck concrete, ksi
- e_g = distance between the centers of gravity of the girder and deck, inches
- A = non-composite area of a single I-girder, in^2
- I = non-composite moment of inertia of a single I-girder, in^4
- N_L = number of loaded lanes
- N_B = number of beams (girders) in the bridge cross section

DOI: 10.1201/9781003265467-5

Note that the units specified in the definitions must be used in the equations. Any necessary unit conversion factors have already been incorporated into the equations.

The value of L to be used in the equations depends on the parameter being calculated, as noted below.

- For positive moments, use L for the span under consideration.
- For negative moment near interior supports between points of contraflexure, use the average of the adjacent spans for L.
- For negative moments other than near interior supports, use L for the span under consideration.
- For shear, use L for the span under consideration.
- For reactions at exterior supports (typically abutments or piers with expansion joints), use L for the span under consideration.
- For reactions at continuous interior supports (piers), use the average of adjacent span lengths for L.

5.1 AASHTO EQUATIONS

The focus of material presented here is on concrete and steel I-girders. For box girders and other types of superstructures, the reader is referred to Section 4.6.2.2 of the AASHTO LRFD BDS (AASHTO, 2020).

The distribution factors in the AASHTO equations have the multi-presence factor, m, built in. There is no need to apply m to a distribution factor calculated from the AASHTO equations. The symbol for the distribution factor in AASHTO is g. The equations listed in AASHTO, and here, define mg.

Since the multi-presence factor is incorporated into the AASHTO equations, it is necessary to divide an equation-based, single-lane distribution factor by $m = 1.2$ for assessment of Fatigue limit state response (lever rule and rigid cross-section methods may also be appropriate for Fatigue distribution factors). The multi-presence factor is not to be applied for the Fatigue limit state.

For prestressed I-girders, the live load distribution factors for moment in an interior girder are given by Equations 5.1 and 5.2. The equations apply for four or more girders, provided each of the following conditions is satisfied:

$$3.5 \leq S \leq 16.0$$

$$4.5 \leq t_s \leq 12.0$$

$$20 \leq L \leq 240$$

$$10,000 \leq K_g \leq 7,000,000$$

$$mg = 0.06 + \left(\frac{S}{14}\right)^{0.40} \left(\frac{S}{L}\right)^{0.30} \left(\frac{K_g}{12Lt_s^3}\right)^{0.10}, \quad \text{1-lane loaded} \qquad (5.1)$$

$$mg = 0.075 + \left(\frac{S}{9.5}\right)^{0.60}\left(\frac{S}{L}\right)^{0.20}\left(\frac{K_g}{12Lt_s^3}\right)^{0.10}, \quad \text{multiple-lanes} \tag{5.2}$$

$$K_g = n\left(I + Ae_g^2\right) \tag{5.3}$$

$$n = \frac{E_B}{E_D} = \frac{1{,}820\sqrt{f'_{cB}}}{1{,}820\sqrt{f'_{cD}}} = \sqrt{\frac{f'_{cB}}{f'_{cD}}} \tag{5.4}$$

With only three girders, the live load distribution factor for moment in an interior girder is to be taken as the lesser value of that derived from Equation 5.1, Equation 5.2, or the lever rule (Section 5.2).

For prestressed I-girders with multiple lanes loaded, the live load distribution factor for moment in an exterior girder is given by Equation 5.6, based on the value of e from Equation 5.5. For I-girders, the distance, d_e, in Equation 5.5 is defined as the horizontal distance, in feet, from the centerline of the exterior I-girder to the face of the curb or barrier. The equations apply for four or more girders with multiple lanes loaded and with $-1.0 \le d_e \le 5.5$.

$$e = 0.77 + \frac{d_e}{9.1} \tag{5.5}$$

$$mg = e\left(mg\right)_{\text{interior}} \tag{5.6}$$

With only one lane loaded, the distribution factor for exterior girder moment is determined by the lever rule. With only three girders, the live load distribution factor for moment in an exterior girder is to be taken as the lesser of the values obtained from Equation 5.6 or the lever rule.

For prestressed I-girders, the live load distribution factor for shear in an interior girder is given by Equations 5.7 and 5.8. The equations apply for four or more girders with:

$$3.5 \le S \le 16.0$$

$$4.5 \le t_s \le 12.0$$

$$20 \le L \le 240$$

$$mg = 0.36 + \frac{S}{25}, \quad \text{1-lane loaded} \tag{5.7}$$

$$mg = 0.2 + \frac{S}{12} - \left(\frac{S}{35}\right)^{2.0}, \quad \text{multiple-lanes loaded} \tag{5.8}$$

With only three girders, the live load distribution factor for shear in an interior girder is to be calculated using the lever rule, regardless of the number of lanes loaded.

For prestressed I-girders with multiple lanes loaded, the live load distribution factor for shear in an exterior girder is given by Equation 5.10. The equation is applicable with four or more girders, with multiple lanes loaded, and with $-1.0 \le d_e \le 5.5$.

$$e = 0.60 + \frac{d_e}{10} \tag{5.9}$$

$$mg = e\left(mg\right)_{\text{interior}} \tag{5.10}$$

With only one lane loaded, or with only three girders, the live load distribution factor for shear in an exterior girder is to be calculated using the lever rule.

The equations presented for prestressed I-girders are also applicable to steel I-girders with cast-in-place deck. However, with steel girder bridges, the cross-section properties are unknown before design of the girder has been completed, so it is not generally possible to determine K_g from equations. AASHTO provides for an approximate distribution factor to be calculated, using Equation 5.11, when girder properties are not known up front.

$$\left(\frac{K_g}{12Lt_s^3}\right)^{0.10} = \begin{cases} 1.02, & \text{steel I-girders} \\ 1.09, & \text{concrete I-girders} \end{cases} \tag{5.11}$$

The final design, once the cross-sectional properties are known, should include calculation of (a) the stiffness parameter, K_g, from Equation 5.3, (b) the new distribution factor with modified K_g, and (c) subsequent re-calculation of live load moments, shears, reactions, and displacements if the modified distribution factor differs significantly from that assumed originally.

Skew effects may be accounted for in live load distribution by the AASHTO equations. Skew tends to decrease moment and increase shear in exterior girders. Equation 5.12 gives the amplification factor for girder shear in bridges with skewed supports, applicable only to shear in the exterior girder at the obtuse corner. The correction factor, CF, is to be applied directly to the calculated distribution factor. For other skew effects, refer to Section 4.6.2.2 of the AASHTO LRFD BDS. Equation 5.12 is applicable subject to the following limitations:

$$0° \le \theta \le 60°$$

$$3.5 \le S \le 16.0$$

$$20 \le L \le 240$$

$$N_B \ge 4$$

$$CF = 1.0 + 0.20\left(\frac{12.0Lt_s^3}{K_g}\right)^{0.3}\tan\theta \qquad (5.12)$$

5.2 THE LEVER RULE

The lever rule is based on a simplifying assumption that the exterior girder and first interior girder act as a propped beam. Each wheel line of the design truck carries one-half of the load (for a straight bridge). The assumption is illustrated in Figure 5.1. By summing moments abut the first interior girder, the reaction at the exterior girder corresponds to the lever rule-based distribution factor.

When a distribution factor is calculated using the lever rule, the multi-presence factor should be applied. An exception is the calculation of distribution factors for the Fatigue limit state, for which multi-presence factors are never to be applied. The Fatigue loading is defined as a single truck in a single lane, with no added uniform live load (as opposed to the live load definition for other limit states).

The distance, x, in Figure 5.1, is taken to be no less than 2 feet for distribution factor calculations. For deck design, the distance x is taken to be no less than 1 foot.

When the bridge is curved in plan, recall that centrifugal forces result in an unequal distribution of truck weight between the inside and outside wheel lines (see Section 2.4, this book). Although often ignored in distribution factor calculations, it may be advisable to consider this unequal distribution of truck weight in the lever-rule analysis for bridges with significant curvature.

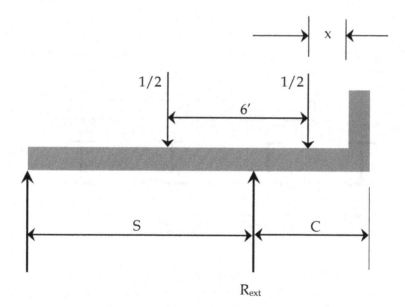

FIGURE 5.1 Live load distribution factor by the lever rule.

5.3   RIGID CROSS-SECTION METHOD

Application of the rigid cross-section method of live load distribution is often necessary for steel girder bridges or any bridge with a cross section which may be relatively rigid, in a torsional manner, through the use of cross-frame-like elements. In
Section 4.6.2.2.2d, the AASHTO LRFD BDS specifies that, for exterior girders in
beam-slab bridges with diaphragms or cross-frames, the live load distribution factor
should also be computed using this method. So, even with concrete girders, if rigid
intermediate diaphragms are present in the bridge, the cross-section method should
be applied. Implicit in the method is the assumption that the cross section rotates
as a rigid body. The live load distribution factor of the rigid cross-section method is
defined by Equation 5.13. N_L is the number of loaded lanes, and N_B is the number of
beams (girders) in the cross section. The distance from the centerline of the bridge
to the center of a lane is the eccentricity, e, for that lane. This is not to be confused
with the adjustment factor, e, used in determining live load distribution factors for
exterior girders, using the AASHTO equations (see Equations 5.5 and 5.9). The distance from the centerline of the bridge to a girder centerline is the parameter, X, for
that girder. The multi-presence factor, m, needs to be incorporated when performing
a rigid cross-section analysis. Lane locations must be determined to maximize each
girder individually (Figure 5.2).

$$mg = m \left[\frac{N_L}{N_B} + \frac{X_{ext} \Sigma e}{\Sigma X^2} \right]$$ (5.13)

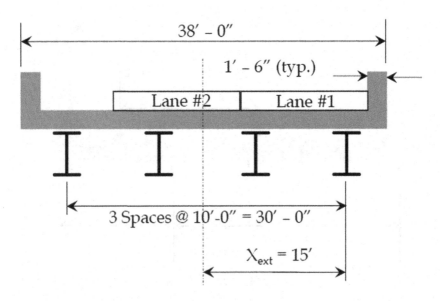

FIGURE 5.2   Live load distribution factor by the rigid cross-section method.

As is the case with lever-rule calculations, it may be advisable to incorporate unequal weight distribution between the inner and outer wheel lines due to centrifugal forces in bridges with plan curvature when using the rigid cross-section method.

5.4   SOLVED PROBLEMS

Problem 5.1

For the steel girder option of the Project Bridge, calculate the live load distribution factors for shear and moment for an exterior girder using all three available methods. Use the following parameters:
- $S = 9$ ft 3 inches
- $C = 3$ ft 4 inches
- W40×215 Girders, $A = 63.5$ in^2, $I = 16,700$ in^4
- $f'_c = 4,000$ psi for the deck

Problem 5.2

For the concrete girder option of the Project Bridge, calculate the live load distribution factors for shear and moment for both interior and exterior girders using all applicable methods. Use the following parameters:
- $S = 9$ ft 3 inches
- $C = 3$ ft 4 inches
- BT-54 Girders, $A = 659$ in^2, $I = 268,077$ in^4
- $f'_c = 4,000$ psi for the deck, $f'_c = 9,000$ psi for the girders

Problem 5.3

A steel I-girder bridge is 77-ft wide with 1-ft wide parapets on both edges. Girder spacing is 10 ft and eight girders are used in the cross section. Determine the exterior girder live load distribution factor by the rigid cross-section method for all applicable loaded land cases. In practice, all other applicable methods (AASHTO equations and lever-rule method) would have to be determined as well to establish the controlling conditions. The purpose of this problem is to fully illustrate the rigid cross-section method.

Problem 5.4

Using the results from Problem 5.1, establish a design moving load case for (a) the Fatigue limit state and (b) for all other limit states. Include any necessary multi-presence and impact factors, but do not include load factors.

PROBLEM 5.1	TEH	1/3

$S = 9'-3''$ $C = 3'-4''$

$W40 \times 215$, $A = 63.5 \text{ in}^2$, $I = 16,700 \text{ in}^4$

$f'_c = 4,000 \text{ psi}$ $e_g = \frac{39}{2} + 2 + \frac{8.25}{2} = 25.6''$

$n = \dfrac{29,000}{1,820\sqrt{4}} = 7.97$

a) AASHTO Equations, Exterior Girders

$d_e = 3'4'' - 1'2\frac{1}{2}'' = 2.125'$

$(mg)_{M-ext} = e \, (mg)_{M-int}$

$K_g = 7.97 \, (16,700 + 63.5 \times 25.6^2)$

$\Rightarrow K_g = 465,421 \text{ in}^4$

$\left(\dfrac{K_g}{12 L t_s^3}\right)^{0.10} = \left(\dfrac{465,421}{12 \times 90 \times 8.25^3}\right)^{0.1}$

$= 0.974$

$(mg)_{M-int} = 0.075 + \left(\dfrac{9.25}{90}\right)^{.2} \left(\dfrac{9.25}{9.5}\right)^{.6} (0.974)$

$= 0.683$ lanes/girder
        (multiple lanes)

$e = 0.77 + 2.125/9.1 = 1.0035$

$(mg)_{M-ext} = 1.0035 \times 0.683$

$= 0.686$ lanes/girder

PROBLEM 5.1	TEH	2/3

$$(mg)_{V-EXT} = e \, (mg)_{V-Int}$$

$$e = .6 + 2.125/10 = 0.8125$$

$$(mg)_{V-Int} = 0.2 + \frac{9.25}{12} - \left(\frac{9.25}{35}\right)^2$$

$$= 0.901 \text{ lanes/girder}$$

$$(mg)_{V-EXT} = 0.8125 \times 0.901$$

$$= 0.732 \text{ lanes/girder}$$

b) Lever Rule

$$(mg)_{LR} = 1.2 \left(\frac{1}{2}\right)\left(\frac{3.37 + 9.37}{9.25}\right)$$

$$\Rightarrow (mg)_{LR} = 0.826 \text{ lanes/girder}$$

c) Rigid Cross-Section, 2 lanes max

lane 1 centered 6.21' from edge

$\Rightarrow e_1 = 11.00'$ from $\cancel{L}$ bridge

$e_2 = 11 - 12 = -1.00$ ft

$\Sigma_i X^2 = 2(4.625^2 + 13.875^2)$

$= 427.8$

1-Lane loaded, $m = 1.2$

$\Sigma_i e = 11.00$

$(mg)_{RXS} = 1.2 \left[\frac{1}{4} + \frac{11 \times 13.875}{427.8} \right]$

$= 0.728$ lanes/girder

2-lanes loaded, $m = 1.0$

$\Sigma_i e = 11 - 1 = 10.00'$

$(mg)_{RXS} = 1.0 \left[\frac{2}{4} + \frac{10 \times 13.875}{427.8} \right]$

$= 0.824$ lanes/girder

	M	V
Eq'ns	0.686	0.732
Lever Rule	0.826	0.826
RX S	0.824	0.824
Use:	0.824	0.824

For Fatigue, use the lever rule without multi-presence factors:

$$(mg)_{Fat} = 0.826/1.2 = 0.688 \text{ lanes per girder}$$

PROBLEM 5.2	TEH	1/3

Use a) AASHTO Equations
     b) Lever rule

$S = 9.25'$ $d_e = 3.33' - 1.21' = 2.12'$

$e_g = (54 - 27.63) + 2 + \dfrac{8.25}{2}$

$\Rightarrow e = 32.5''$

$n = \dfrac{1,820\sqrt{9}}{1,820\sqrt{4}} = 1.50$

$K_g = 1.50\,(268,077 + 659 \times 32.5^2)$

$\quad = 1,446,219 \text{ in}^4$

$\left(\dfrac{K_g}{12 L t_s^3}\right)^{0.1} = \left(\dfrac{1,446,219}{12 \times 90 \times 8.25^3}\right)^{0.1} = 1.091$

a) AASHTO Equations, Interior Girder
   1-lane loaded, moment

$(mg)_{M\text{-int}} = .06 + \left(\dfrac{9.25}{14}\right)^{.4}\left(\dfrac{9.25}{90}\right)^{.3}(1.091)$

$\quad = 0.527 \text{ lanes/girder}$

Multiple lanes, moment

$(mg)_{M\text{-int}} = .075 + \left(\dfrac{9.25}{9.5}\right)^{.6}\left(\dfrac{9.25}{90}\right)^{.2}(1.091)$

$\quad = 0.756 \text{ lanes/girder}$

PROBLEM 5.2	TEH	2/3

1-Lane Loaded, Shear.

$$(mg)_{V-Int} = .36 + \frac{9.25}{25} = 0.730$$

Multiple Lanes, Shear

$$(mg)_{V-Int} = .2 + \frac{9.25}{12} - \left(\frac{9.25}{35}\right)^2 = 0.901$$

b) AASHTO Equations, Exterior Girder

1-lane loaded, Lever rule

$$(mg)_{LR} = 1.2 \left[\frac{1}{2}\left(\frac{3.37 + 9.37}{9.25}\right)\right]$$

$$= 0.826 \text{ lanes/girder}$$

PROBLEM 5.2	TEH	3/3

Multiple lanes loaded, moment

$$e = 0.77 + 2.12/9.1$$
$$= 1.003$$

$$(mg)_{M-EXT} = 1.003 \times 0.756$$
$$= 0.758 \text{ lanes /girder}$$

Multiple lanes loaded, shear

$$(mg)_{V-EXT} = (.6 + 2.12/10)(0.901)$$
$$= 0.732 \text{ lanes /girder}$$

	M	V
Interior	0.756	0.901
Exterior	0.826	0.826

For fatigue, use lever rule
with $m = 1.2$ ∴

$$(mg)_{fat} = 0.826/1.2 = 0.688$$

PROBLEM 5.3	TEH	1/2

$$N_L \text{ Max} = \text{Int}\left[(77-2)/12\right] = 6 \text{ Lanes}$$

7 SPACES @ 10' (Girders)

$$\sum_1^4 x^2 = 2\left(5^2 + 15^2 + 25^2 + 35^2\right)$$

$$= 4,200 \text{ ft}^2$$

1-Loaded Lane. $e_1 = \dfrac{77}{2} - 1 - 5 = 32.5 \text{ ft}$
(m=1.2)

$$(mg)_{RXS} = 1.2\left(\frac{1}{8} + \frac{32.5 \times 35}{4,200}\right) = 0.475$$

2-Loaded Lanes. $e_2 = 32.5 - 12 = 20.5'$
(m=1.0)
$$\sum_1^2 e = 32.5 + 20.5 = 53$$

$$(mg)_{RXS} = 1.0\left(\frac{2}{8} + \frac{53 \times 35}{4,200}\right) = 0.692$$

3-Loaded Lanes. $e_3 = 20.5 - 12 = 8.5'$
$$\sum_1^3 e = 53 + 8.5 = 61.5'$$

$$m = 0.85$$

$$(mg)_{RXS} = 0.85\left(\frac{3}{8} + \frac{61.5 \times 35}{4,200}\right) = 0.754$$

PROBLEM 5.3	TEH	2/2

4. Loaded Lanes. $e_4 = 8.5 - 12 = -3.5'$

$$\sum_1^4 e = 61.5 - 3.5 = 58'$$

$$m = 0.65$$

$$(mg)_{RXS} = 0.65 \left(\frac{4}{8} + \frac{58 \times 35}{4,200} \right) = 0.639$$

5 - Loaded Lanes. $e_5 = -3.5 - 12 = -15.5$

$$\sum_1^5 e = 58 - 15.5 = 42.5$$

$$m = 0.65$$

$$(mg)_{RXS} = 0.65 \left(\frac{5}{8} + \frac{42.5 \times 35}{4,200} \right) = 0.636$$

6 - Loaded Lanes. $e_6 = -15.5 - 12 = -27.5$

$$\sum_1^6 e = 42.5 - 27.5 = 15.00'$$

$$m = 0.65$$

$$(mg)_{RXS} = 0.65 \left(\frac{6}{8} + \frac{15 \times 35}{4,200} \right) = 0.569$$

Worst Case for rigid cross-section method is with 3-loaded lanes

$$(mg)_{RXS} = 0.754 \text{ lanes/girder}$$

PROBLEM 5.4	TEH	1/2

From Problem 5-1,

mg = 0.824 lanes per girder
for all limit states
except fatigue

mg = 0.688 lanes per girder
for the fatigue limit
state.

Fatigue Limit State

mg = 0.688 lanes/girder
IM = 0.15, applied to
truck
no lane loading
front axle, W_1: $8^K \times 1.15 \times .688$

= 6.33 kips

middle & rear axles:
$W_2 = W_3 = 32^K \times 1.15 \times .688$

= 25.32 kips

All other Limit States.

mg = 0.824 lanes/girder
IM = 0.33, truck only
Lane load included

| PROBLEM 5.4 | TEH | 2/2 |

front axle,

$$W_1 = 8^k \times 1.33 \times 0.824 = 8.77k$$

middle & rear axles;

$$W_2 = W_3 = 32^k \times 1.33 \times 0.824$$
$$= 35.07 k$$

uniform load,

$$p = 0.64 \text{ KLF} \times 0.824$$
$$= 0.527 \text{ KLF}$$

Fatigue Case:

6.33^k 25.32^k 25.32^k

14'   30'

All other Limit States:

8.77^k 35.07^k 35.07^k

14'   Varies 14' to 30'

0.527 KLF

5.5   EXERCISES

E5.1

A 284-ft long, two-span bridge consists of four prestressed concrete BT-72 I-girders spaced 8 ft 3 inches apart. The bridge is 31 ft 3 inches wide and has equal span lengths. The inside face of the parapet is 1 foot from the outer edge of the bridge deck. Girder concrete has a minimum specified 28-day concrete strength of 10 ksi, while the deck uses 4 ksi concrete. Deck thickness is 8.25 inches. The haunch (distance from top of girder to bottom of deck) is 2 inches. The properties may be assumed constant for the entire bridge length. For the Service and Strength limit states, determine:
- distribution factors for moment and shear for an interior girder
- distribution factors for moment and shear for an exterior girder

E5.2

For the bridge of Exercise E5.1, determine the appropriate distribution factors for shear and moment at the Fatigue limit state for (a) an interior girder and (b) an exterior girder. Use a continuous beam computer model to generate shear and moment envelopes for the controlling girder for the Fatigue limit state.

E5.3

A steel I-girder bridge is 57 feet wide. Five girders spaced 12 feet apart are used for the bridge cross-section. The parapet width is 1 ft 6 inches at the base. Determine the live load distribution factors for an exterior girder using the rigid cross-section method. Consider all possible numbers of loaded lanes.

E5.4

For the steel superstructure option of the Project Bridge, determine the live load distribution factors for moment and shear in an interior girder. Model the continuous spans in VisualAnalysis and create moving load combinations to determine shear and moment envelopes for LL+IM corresponding to the Fatigue limit state (IM = 15%) and all other limit states (IM = 33%). Plot the envelopes. Verify using the National Steel Bridge Alliance (NSBA) LRFD Simon software. Use girder spacing $S = 9$ ft 3 inches and W40 × 215 girders. Deck concrete f'_c is equal to 4,000 psi.

E5.5

A 57-ft 6-inch-wide bridge is curved in plan with centerline radius of curvature equal to 750 ft. Parapet width is 1 ft at the base. Girder spacing, $S = 12$ feet and the overhang dimension, $C = 4$ ft 9 inches. Design speed is 50 mph. Incorporate the centrifugal force effect on wheel line distribution and compute live load distribution factors using (a) the lever-rule method and (b) the rigid cross-section method. Superelevation is 0.08 ft/ft.

6 Steel Welded Plate I-Girders

Details to be developed and designed for steel girder bridges include each of the following:

- girder transitions (flange width and thickness; web depth and thickness)
- bolted field splices
- cross-frame layout
- cross-frame details
- bearing stiffeners
- shear stiffeners
- cross-frame connection stiffeners
- shear studs
- camber diagrams
- web-to-flange weld
- girder stability (lateral bracing; torsional bracing)
- bearings
- pouring sequence

The focus of the material presented in this chapter is on the Strength and Fatigue limit states. There are many other requirements for steel I-girder bridges, including those for Constructability and Service limit states. For a detailed coverage of design information not covered here for steel I-girders, refer to the AASHTO LRFD BDS, Section 6.10 (AASHTO, 2020).

To qualify as a compact section in positive flexure at the Strength limit state in the AASHTO LRFD BDS, a steel I-girder must satisfy each of Equations 6.1, 6.2, and 6.3. Parameters in the equations include the depth of the web in compression (D_{cp}), the web thickness (t_w), the total web depth (D), Young's modulus ($E = 29{,}000$ ksi), and the yield strength of the compression flange (F_{yc}).

$$F_y \leq 70 \text{ ksi for flanges} \tag{6.1}$$

$$\frac{2D_{cp}}{t_w} \leq 3.76\sqrt{\frac{E}{F_{yc}}} \tag{6.2}$$

$$\frac{D}{t_w} \leq 150 \tag{6.3}$$

DOI: 10.1201/9781003265467-6

D_{cp} is the web depth in compression for the plastic stress distribution. Section proportion limits are summarized for I-girders next.

For the webs of I-girders:

- $D/t_w \leq 150$ with no longitudinal stiffeners
- $D/t_w \leq 300$ with longitudinal stiffeners

For the flanges of I-girders:

- flange width (b_f)/thickness (t_f) limit: $b_f \leq 24\, t_f$
- flange width/web depth limit: $b_f \geq D/6$
- flange thickness/web thickness limit: $t_f \geq 1.1\, t_w$
- $0.10 \leq I_{yc}/I_{yt} \leq 10$
- $I_{yc} = (t_f)(b_f)^3/12$, compression flange
- $I_{yt} = (t_f)(b_f)^3/12$, tension flange

Stiffened webs must satisfy Equation 6.4, or there is a penalty on shear resistance. Subscripts "c" and "t" refer to the compression and tension flanges, respectively.

$$\frac{2Dt_w}{b_{fc}t_{fc} + b_{ft}t_{ft}} \leq 2.5 \tag{6.4}$$

6.1 FLEXURAL RESISTANCE AT THE STRENGTH LIMIT STATE

For complete coverage of flexural resistance at each limit state for all types of steel girders, refer to the AASHTO LRFD BDS, Section 6.10 for I-girders and Section 6.11 for box girders. Discussions in this chapter are limited to (a) positive flexural resistance of composite, compact sections at the Strength limit state, Section 6.1.1, (b) positive flexural resistance of non-compact composite sections at the Strength limit state in Section 6.1.2, and (c) negative flexure at the Strength limit state in Section 6.1.3.

6.1.1 COMPOSITE COMPACT SECTIONS IN POSITIVE FLEXURE

Nominal, positive flexural resistance of compact sections is determined by Equation 6.5a. D_p is the distance from the extreme compression fiber to the plastic neutral axis (PNA). D_t is the total composite girder depth, including the deck. In continuous span homogeneous girders, the moment resistance, M_n, is further limited to $1.3 \times M_y$, as indicated in Equation 6.5a. To find M_y, solve Equation 6.5b for M_{AD}, using factored moments M_{D1} and M_{D2}. M_{D1} is the factored dead load moment carried by the girder alone (*NC* denotes "non-composite"). M_{D2} is the factored dead load moment carried by the long-term composite section (*LT* denotes "long-term composite"). Then substitute into Equation 6.5c. Refer to the AASHTO LRFD BDS, Section D6.2, for more information on the computation of M_y. Figure 6.1 will clarify the definitions of the various depths for a composite section in positive flexure (top of the deck in compression).

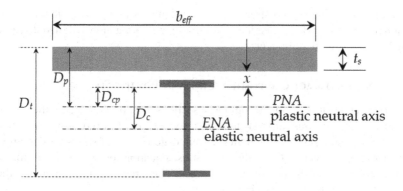

FIGURE 6.1   Composite section in positive flexure.

Refer to Section 6.9 of this chapter for a more detailed discussion on the computation of plastic moment resistance, M_p, for non-composite and composite steel I-girders.

$$M_n = \begin{cases} M_p, & \text{if } \dfrac{D_p}{D_t} \le 0.1 \\[2mm] M_p\left(1.07 - 0.7\dfrac{D_p}{D_t}\right), & \text{if } 0.10 < \dfrac{D_p}{D_t} \le 0.42 \\[2mm] 0, & \text{if } \dfrac{D_p}{D_t} > 0.42 \,(\text{not permitted}) \\[2mm] \le 1.3 R_h M_y \,(\text{continuous spans}) \end{cases} \tag{6.5a}$$

$$F_{yf} = \frac{M_{D1}}{S_{NC}} + \frac{M_{D2}}{S_{LT}} + \frac{M_{AD}}{S_{ST}} \tag{6.5b}$$

$$M_y = M_{D1} + M_{D2} + M_{AD} \tag{6.5c}$$

$$M_u + \frac{1}{3} f_l S_{xt} \le \phi_f M_n \tag{6.5d}$$

The required design resistance and available design resistance must satisfy Equation 6.5d. The lateral flange bending stress, f_l, may be determined according to Equation 6.9 and as discussed below. The section modulus, S_{xt}, is that for the tension flange and is equal to M_{yt}/F_{yt}. Refer to AASHTO LRFD BDS, Article D6.2, for additional information on the computation of M_{yt}.

Positive moment sections which are composite in the final conditions require assessment at various stages of construction. Significant loading due to wet concrete and construction live loads are a critical consideration. Stability assessment prior to the deck reaching design strength is particularly critical. For the completed structure, the deck provides lateral bracing of the top flange under typical circumstances. Prior to the deck reaching design strength, the top flange is more susceptible to lateral-torsional buckling. Refer to Section 6.7 of this chapter for a more detailed discussion of stability bracing requirements.

6.1.2 NON-COMPACT COMPOSITE SECTIONS IN POSITIVE FLEXURE

Non-compact, positive moment, composite sections must be assessed for the Strength limit state based on stress analysis for the various stages of the construction process, namely non-composite (NC), long-term composite (LT), and short-term composite (ST) properties. Equation 6.6 is the basic stress equation for flexure due to loads at the various stages of the construction process. The stress, f_{bu}, is computed without contribution from lateral flange bending.

For non-compact, positive moment, composite sections, Equation 6.7 must be satisfied for the compression flange and Equation 6.8 must be satisfied for the tension flange, both under Strength limit state load factors. The first-order lateral flange bending stress, f_l, may be estimated using Equation 6.9, with consistent units, for straight I-girders under wind-induced moment, M_w. The compression flange nominal resistance, F_{nc}, and the tension flange nominal resistance, F_{nt}, are given by Equation 6.10 and 6.11, respectively. For homogeneous (non-hybrid) girders, the hybrid factor, R_h, is equal to 1.00. For hybrid girders, refer to Section 6.10.1.10.1 of the AASHTO LRFD BDS. The web load-shedding factor, R_b, is also equal to 1.00 for composite sections in positive flexure, with D/t no more than 150. For other cases, refer to Section 6.10.1.10.2 of the AASHTO LRFD BDS.

$$f_{bu} = \pm\left(\frac{M_{NC}}{S_{NC}}\right) \pm \left(\frac{M_{LT}}{S_{LT}}\right) \pm \left(\frac{M_{ST}}{S_{ST}}\right) \tag{6.6}$$

$$f_{bu} \leq \phi_f F_{nc} \tag{6.7}$$

$$f_{bu} + \frac{1}{3}f_l \leq \phi_f F_{nt} \tag{6.8}$$

$$f_l = \frac{6M_w}{tb^2} \tag{6.9}$$

$$F_{nc} = R_b R_h F_{yc} \tag{6.10}$$

$$F_{nt} = R_h F_{yt} \tag{6.11}$$

For composite deck bridges, the wind load on the lower half of the outside I-girder may be assumed to be carried by the bottom flange in lateral bending. The wind load on the upper half of the I-girder may be assumed to be transmitted directly to the concrete deck. Frame action of cross-frames may be used, essentially treating the wind-loaded flange as a continuous beam supported at cross-frame locations. As a result, the wind moment, M_w, may be taken to be equal to that given by Equation 6.12. The load, W, is the wind load per foot-length acting on the flange, and may be calculated using Equation 6.13 for the composite deck conditions previously stated. The girder depth is d, and P_D is the design wind pressure at the appropriate Strength limit state. The cross-frame spacing is L_b.

$$M_w = \frac{WL_b^2}{10} \tag{6.12}$$

$$W = \frac{\eta_i \gamma P_D d}{2} \tag{6.13}$$

Flange lateral bending stress, f_l, may be taken to be equal to the first-order value from Equation 6.9, provided Equation 6.14 is satisfied. Otherwise, the value calculated from Equation 6.9 requires amplification to account for second-order effects. Refer to the AASHTO LRFD BDS, Section 6.10.1.6. See also AASHTO LRFD BDS, Section 6.10.8.2.3, for a detailed calculation of the moment gradient factor, C_b, in cases where a value other than $C_b = 1.00$ may be used.

$$L_b \leq 1.2 L_p \sqrt{\frac{C_b R_b}{f_{bu} / F_{yc}}} \tag{6.14}$$

Factored lateral flange bending stress, f_l, at the Strength limit state must never exceed $0.6 F_{yf}$.

For computation of lateral flange bending stresses in curved girders, the reader is referred to Chapter 6 of the AASHTO LRFD BDS (AASHTO, 2020).

6.1.3 NEGATIVE FLEXURE AND NON-COMPOSITE SECTIONS

For sections in negative flexure and for non-composite sections, Equation 6.15a defines the compression flange resistance to local buckling. Equation 6.15b defines the compression flange resistance to lateral torsional buckling. The web load-shedding factor, R_b, is given by Equation 6.24b for cross sections in negative flexure and non-composite cross sections. F_{nc} is never to be taken to be greater than $R_b R_h F_{yc}$. The flange stresses at the Strength limit state must satisfy Equations 6.15g (for the compression flange) and 6.15h (for the tension flange). Alternative methods from Appendix A6 of the AASHTO LRFD BDS may be followed in lieu of those discussed here for non-composite sections and sections in negative flexure. Further limitations on sections in negative flexure and non-composite sections are given in Equations 6.20 and 6.21. The moment gradient factor, C_b, may conservatively be taken to be equal to 1.0, representing an unbraced segment with constant moment over the entire segment. Refer to the AASHTO LRFD BDS, Article 6.10.8.2.3, for detailed calculation of C_b to justify values greater than 1.0.

Equation 6.21 for web slenderness must be satisfied in order to permit the alternative Appendix A6 method for evaluating the cross section. Otherwise, the stress analysis and limits presented here for non-compact sections are applicable.

$$F_{nc} = \begin{cases} R_b R_h F_{yc}, & \text{if } \lambda_f \leq \lambda_{pf} \\ \left[1 - \left(1 - \dfrac{F_{yr}}{R_h F_{yc}} \right) \left(\dfrac{\lambda_f - \lambda_{pf}}{\lambda_{rf} - \lambda_{pf}} \right) \right] R_b R_h F_{yc}, & \text{if } \lambda_f > \lambda_{pf} \end{cases} \quad (6.15a)$$

$$F_{nc} = \begin{cases} R_b R_h F_{yc}, & \text{if } L_b \leq L_p \\ C_b \left[1 - \left(1 - \dfrac{F_{yr}}{R_h F_{yc}} \right) \left(\dfrac{L_b - L_p}{L_r - L_p} \right) \right] R_b R_h F_{yc}, & \text{if } L_p < L_b < L_r \quad (6.15b) \\ F_{cr} \leq R_b R_h F_{yc}, & \text{if } L_b > L_r \end{cases}$$

$$L_p = 1.0 r_t \sqrt{\frac{E}{F_{yc}}} \tag{6.15c}$$

$$L_r = \pi r_t \sqrt{\frac{E}{F_{yr}}} \tag{6.15d}$$

$$r_t = \frac{b_{fc}}{\sqrt{12 \left(1 + \dfrac{1}{3} \dfrac{D_c t_w}{b_{fc} t_{fc}} \right)}} \tag{6.15e}$$

$$F_{cr} = \frac{C_b R_b \pi^2 E}{\left(L_b \middle/ r_t \right)^2} \tag{6.15f}$$

$$f_{bu} + \frac{1}{3} f_l \leq \phi_f F_{nc} \tag{6.15g}$$

$$f_{bu} + \frac{1}{3} f_l \leq \phi_f R_h F_{yt} \tag{6.15h}$$

$$\lambda_f = \frac{b_{fc}}{2 t_{fc}} \tag{6.16}$$

$$\lambda_{pf} = 0.38 \sqrt{\frac{E}{F_{yc}}} \tag{6.17}$$

$$\lambda_{rf} = 0.56 \sqrt{\frac{E}{F_{yc}}} \tag{6.18}$$

$$F_{yr} = 0.70 F_{yc}, \text{ for homogeneous girders} \qquad (6.19)$$

$$F_{yf} \leq 70 \, \text{ksi} \qquad (6.20)$$

$$\frac{2D_c}{t_w} \leq \lambda_{rw} \qquad (6.21)$$

$$\frac{I_{yc}}{I_{yt}} \geq 0.30 \qquad (6.22)$$

$$\lambda_{rw} = 4.6 \sqrt{\frac{E}{F_{yc}}} \leq \left(3.1 + \frac{5.0}{a_{wc}}\right)\sqrt{\frac{E}{F_{yc}}} \leq 5.7\sqrt{\frac{E}{F_{yc}}} \qquad (6.23)$$

$$a_{wc} = \frac{2D_c t_w}{b_{fc} t_{fc}} \qquad (6.24a)$$

$$R_b = 1.0 - \frac{a_{wc}}{1200 + 300 a_{wc}}\left(\frac{2D_c}{t_w} - \lambda_{rw}\right) \qquad (6.24b)$$

- D_c = depth of web in compression in the elastic stress range, inches
- b_{fc} = compression flange width, inches
- t_{fc} = compression flange thickness, inches
- b_{ft} = tension flange width, inches
- t_{ft} = tension flange thickness, inches
- F_{yc} = compression flange yield stress, ksi
- t_w = web thickness, inches
- d = total girder depth, inches
- $I_{yc} = t_{fc}(b_{fc})^3/12$
- $I_{yt} = t_{ft}(b_{ft})^3/12$

The material presented here is applicable for straight I-girder bridges, with no skew, and with cross-frame lines continuous across the cross section of the bridge.

At locations with holes in the tension flange (at field splices, for example), for the Strength limit state and for constructability, the computed flange stress is limited to $0.84(A_n/A_g)F_u \leq F_y$. Refer to AASHTO LRFD BDS, Section 6.10.1.8, for a detailed discussion.

6.2   SHEAR RESISTANCE

The shear resistance of stiffened I-girders is given by Equation 6.25. The shear resistance of unstiffened panels is computed using the equation for no tension field action (TFA) and $k = 5$. The resistance factor for shear in steel I-girders is $\phi_v = 1.00$. The clear distance between transverse stiffeners is d_o. The plastic shear, $V_P = 0.58 F_{yw} D t_w$.

$$V_n = \begin{cases} CV_p & \text{for end panels with } d_o \leq 1.5D, \text{ No } TFA \\ V_p \left[C + \dfrac{0.87(1-C)}{\sqrt{1 + \left(\dfrac{d_o}{D} \right)^2}} \right] & \text{for intermediate panels, } TFA \end{cases} \tag{6.25}$$

$$k = 5 + \frac{5}{\left(d_o / D \right)^2} \tag{6.26}$$

$$C = \begin{cases} 1 & \text{if } \dfrac{D}{t_w} \leq 1.12 \sqrt{\dfrac{Ek}{F_{yw}}} \\[2mm] \dfrac{1.12}{D/t_w} \sqrt{\dfrac{Ek}{F_{yw}}} & \text{if } 1.12 \sqrt{\dfrac{Ek}{F_{yw}}} \leq \dfrac{D}{t_w} \leq 1.40 \sqrt{\dfrac{Ek}{F_{yw}}} \\[2mm] \dfrac{1.57}{\left(D/t_w \right)^2} \cdot \dfrac{Ek}{F_{yw}} & \text{if } \dfrac{D}{t_w} \geq 1.40 \sqrt{\dfrac{Ek}{F_{yw}}} \end{cases} \tag{6.27}$$

These equations are applicable to transversely stiffened I-girders. The design requirements for the transverse stiffeners will be presented next. The design requirements for I-girders with longitudinal stiffeners is beyond the scope of this text.

6.3   TRANSVERSE STIFFENER DESIGN

Stiffeners in straight girders not used as connection plates should be tightly fitted or attached at the compression flange but need not be in bearing with the tension flange. Single-sided stiffeners on horizontally curved girders should be attached to both flanges. When pairs of transverse stiffeners are used on horizontally curved girders, they must be fitted tightly or attached to both flanges.

Stiffeners used as connecting plates for diaphragms or cross-frames must be attached to both flanges.

The distance between the end of the web-to-stiffener weld and the near edge of the adjacent web-to-flange weld must be no less than $4t_w$ but cannot exceed the lesser of $6t_w$ or 4.0 inches.

Transverse stiffener dimensions must satisfy Equations 6.28 and 6.29. Properties of transverse stiffeners must satisfy Equation 6.30.

$$b_t \geq 2 + \frac{D}{30} \tag{6.28}$$

$$16t_p \geq b_t \geq \frac{b_f}{4} \tag{6.29}$$

$$I_t \geq \begin{cases} \text{Min}\left[bt_w^3 J, \dfrac{D^4 \rho_t^{1.3}}{40}\left(\dfrac{F_{yw}}{E}\right)^{1.5} \right], \text{ with no } TFA \\[20pt] \dfrac{D^4 \rho_t^{1.3}}{40}\left(\dfrac{F_{yw}}{E}\right)^{1.5}, \text{ with } TFA \end{cases} \tag{6.30}$$

$$J = \frac{2.5}{\left(d_o/D\right)^2} - 2.0 \geq 0.50 \tag{6.31}$$

$$F_{crs} = \frac{0.31E}{\left(b_t/t_p\right)^2} \leq F_{ys} \tag{6.32}$$

$$\rho_t = \text{Max}\left\{\frac{F_{yw}}{F_{crs}}, 1.0\right\} \tag{6.33}$$

$$b = \text{Min}\left\{d_o, D\right\} \tag{6.34}$$

- I_t = moment of inertia of the transverse stiffener taken about the edge in contact with the web for single stiffeners and about the mid-thickness of the web for stiffener pairs, in^4.
- d_o = the smaller spacing on either side of the stiffener, inches
- b_t = stiffener width, inches
- t_p = stiffener thickness, inches
- $F_{ys} = F_y$ for the stiffener, ksi

6.4   BEARING STIFFENER DESIGN

Bearing stiffeners are placed on the webs of built-up I-girder sections at all bearing locations. Bearing stiffeners are comprised of plates or angles welded or bolted to both sides of the web, must extend the full depth of the web, and extend as closely as practical to the outer flange edges. Each bearing stiffener must be finished to bear against the flange through which load is transmitted. This would be the bottom flange in a continuous I-girder at intermediate piers. Bearing stiffeners serving as connection plates must be attached to both flanges of the cross section as specified in AASHTO LRFD BDS, Section 6.6.1.3.1.

Bearing stiffeners are checked for proportion limits, bearing resistance, and axial resistance. For the bearing resistance check, only the portion of the clipped stiffener in contact with the flanges is included as the effective area. For the axial resistance check, the full stiffener width may be used along with a strip of web extending no more than $9t_w$ on each side of the stiffeners to form a column of effective length $KL = 0.75D$.

Equation 6.35 provides the proportion check criterion for bearing stiffeners. Equation 6.36 gives the bearing resistance. Axial resistance is determined using Equation 6.37. "r" is the radius of gyration of the effective column section about the mid-thickness of the web. Bearing resistance and axial resistance must both exceed the Strength limit state reaction.

$$b_t \le 0.48 t_p \sqrt{\frac{E}{F_{ys}}} \tag{6.35}$$

$$\phi_b \left(R_{sb} \right)_n = 1.0 \left(1.4 A_{pn} F_{ys} \right) \tag{6.36}$$

$$\phi_c P_n = 0.95 P_n \tag{6.37}$$

$$P_e = \frac{\pi^2 E}{\left(\frac{KL}{r} \right)^2} A_g \tag{6.38}$$

$$P_o = F_y A_g \tag{6.39}$$

$$P_n = \begin{cases} \left[0.658^{\frac{P_o}{P_e}} \right] P_o, & \text{if } \frac{P_e}{P_o} \ge 0.44 \\[2ex] 0.877 P_e, & \text{if } \frac{P_e}{P_o} < 0.44 \end{cases} \tag{6.40}$$

6.5   FATIGUE DESIGN

Particularly with steel girder bridges, it is critical to consider Fatigue in design. Fatigue is of concern any time cyclic stress exists, with one extreme of the stress range being tension. Fatigue may be described as the (a) initiation and (b) propagation of cracks at stress levels which may be well below yield.

Repeated application of cyclic stress, in terms of stress range, may produce extensive damage in structural elements. Figure 6.2 depicts multiple Fatigue resistance curves defining the number of cycles to failure for various stress ranges and detail categories.

Bridge elements and details are assigned a Fatigue Category. These are represented by the various curves in Figure 6.2. Each Fatigue Category has associated parameters, A and ΔF_{TH}. The various categories in AASHTO LRFD BDS and the associated parameters are summarized in Table 6.1. For those familiar with Fatigue requirements for building design in steel, there are differences between AASHTO for bridges and AISC 360-16 for buildings, and the requirements from the different specifications must not be confused.

Table 6.2 is a small sample of the many details with assigned Fatigue Category found in AASHTO LRFD BDS (AASHTO, 2020).

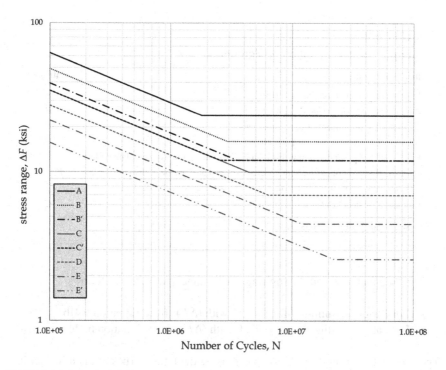

FIGURE 6.2 Typical fatigue curve.

TABLE 6.1
Fatigue Parameters

Category	A, ksi³	ΔF_TH, ksi
A	250×10^8	24.0
B	120×10^8	16.0
B'	61×10^8	12.0
C	44×10^8	10.0
C'	44×10^8	12.0
D	22×10^8	7.0
E	11×10^8	4.5
E'	3.9×10^8	2.6

AASHTO considers two distinct Fatigue limit states:

- Fatigue I, infinite life
- Fatigue II, finite life

For each Fatigue limit state, Equation 6.41 defines the criteria used to assess fatigue resistance for any element, based on actual stress range (Δf) and permissible stress range (ΔF).

TABLE 6.2
Sample Fatigue Category Assignments

Brief Description[a]	Category
Base metal, except uncoated weathering steel	A
Base metal, uncoated weathering steel	B
Base metal at re-entrant corners, copes, cuts, ...	C
Base metal at the net section of open holes	D
Base metal and weld metal in web-to-flange continuous fillet welded plate girders	B
Base metal at ends of partial length welded cover-plates narrower than the flange	E or E'
Base metal at the toe of stiffener-to-flange and stiffener-to-web fillet welds	C'
Base metal at steel headed stud anchors	C

[a] See AASHTO LRFD BDS (AASHTO, 2020) for complete descriptions

The load factor, γ, and the limiting stress range, $(\Delta F)_n$, are as follows:

- For Fatigue I (infinite life), $\gamma = 1.75$ with $(\Delta F)_n$ from Equation 6.41b
- For Fatigue II (finite life), $\gamma = 0.80$ with $(\Delta F)_n$ from Equation 6.41c

The actual stress range, Δf, is computed using the fatigue truck with an impact factor, $IM, = 0.15$, without application of the multi-presence factor (m), and a single-lane live load distribution factor.

The number of cycles, N, expected over the life of the bridge is dependent upon (a) the anticipated 75-year, single-lane, average daily truck traffic ($ADTT_{SL}$), and (b) the number of stress cycles per truck passage, n. The $ADTT_{SL}$ would ideally be based on reliable projections. When such projections are not available, Section 3.6.1 of AASHTO LRFD BDS provides means of estimating the $ADTT_{SL}$ when either the average daily traffic (ADT) or the average daily truck traffic ($ADTT$) is known. Equation 6.41d may be used in such cases. Calculation of the total number of cycles follows from Equation 6.41e.

$$\gamma\left(\Delta f\right) \le \left(\Delta F\right)_n \tag{6.41a}$$

$$\left(\Delta F\right)_n = \left(\Delta F\right)_{TH} \tag{6.41b}$$

$$\left(\Delta F\right)_n = \left(\frac{A}{N}\right)^{1/3} \tag{6.41c}$$

$$ADTT_{SL} = p \times ADTT = fr \times p \times ADT \tag{6.41d}$$

$$N = 365 \times 75 \times n \times ADTT_{SL} \tag{6.41e}$$

The fraction of trucks in traffic, fr, may be estimated as follows:

- Rural interstate, $fr=0.20$
- Urban interstate, $fr=0.15$
- Other rural, $fr=0.15$
- Other urban, $fr=0.10$

The fraction of truck traffic in a single lane may be estimated from:

- 1 lane available to trucks, $p=1.00$
- 2 lanes available to trucks, $p=0.85$
- 3 or more lanes available to trucks, $p=0.80$

The number of stress cycles per truck passage, n, is taken to be equal to 1.5 for a continuous, longitudinal I-girder at regions near interior supports, and 1.0 elsewhere for the I-girder.

6.6 FIELD SPLICE DESIGN

The maximum length which may be shipped to a site varies depending on the quality and geometry of roads, availability of equipment, etc. Usually, 140 ft to 160 ft is a typical maximum length which may be shipped. For steel girder bridges, the field sections are shipped to the site and bolted together in place. Figure 6.3 depicts an example of a field splice detail.

Flange splice filler plates are required any time the flange thickness is not the same for the two sections on opposing sides of the splice centerline.

The National Steel Bridge Alliance (NSBA) provides a free Excel spreadsheet for field splice analysis and design, *NSBA-Splice* (https://www.aisc.org/nsba/design-resources/nsba-splice/).

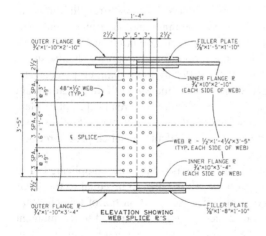
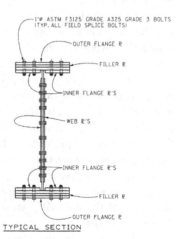

FIGURE 6.3 Steel welded plate girder field Splice.

For extensive design examples by hand calculation, refer to **Bolted Field Splices for Steel Bridge Flexural Members: Overview and Design Examples** (Grubb et al., 2018), also published by the NSBA.

Splice elements include:

- flange splice plates (inner and outer)
- flange bolts
- web splice plates
- web bolts
- filler plates

Bolted field splices are to be designed slip-critical at the Service II limit state.

The flange splice design force, P_{fy}, at the Strength limit state is determined according to Equations 6.42 through 6.44. The controlling flange force for each flange is the smaller of the two values on each side of the splice at said flange. In the 9th edition of AASHTO LRFD BDS, $\phi_u = 0.80$ and $\phi_y = 0.95$.

$$P_{fy} = F_{yf} \cdot A_e \tag{6.42}$$

$$A_e = \frac{\phi_u F_u}{\phi_y F_y} \cdot A_n \leq A_g \tag{6.43}$$

$$A_n = t_f \left[b_f - n d_h \right] \tag{6.44}$$

For standard holes, the diameter of the holes, d_h, is taken to be the bolt diameter plus 1/16 inches for bolt diameters equal to 7/8 inches or less. For larger bolts, the diameter of the holes, d_h, is taken to be equal to the bolt diameter plus 1/8 inch. Unlike the requirements from AISC 360-16 for buildings, no deduction for hole damage is incorporated into AASHTO requirements for net area calculation.

For a flange splice with inner and outer splice plates, P_{fy} at the Strength limit state may be assumed to be divided equally between the inner and outer plates and their connections when the areas of the inner and outer plates do not differ by more than ten percent. In this case, the connections are proportioned assuming double shear. Should the areas of the inner and outer plates differ by more than ten percent, the design force in each splice plate and its connection at the Strength limit state should instead be determined by multiplying P_{fy} by the ratio of the area of the splice plate under consideration to the total area of the inner and outer splice plates. For this case, the connections are proportioned for the maximum calculated splice-plate force acting on a single shear plane.

The determination of web splice design forces is a bit more complicated. The vertical component of web splice design force, V_r, is taken to be equal to the shear resistance of the web, ϕV_n, at the point of splice (see Section 6.2 of this chapter). The Strength limit state moment at the splice location is compared to the moment resistance provided by the flanges. If the Strength limit state moment is greater than the flexural resistance of the flanges, then an additional horizontal force, H_w, must be included in the design to enable the web to carry the additional moment. It is, therefore, necessary to establish a means of determining the moment resistance provided by the flanges.

For composite sections in positive bending (compression in the deck), a moment arm, A, is taken to be equal to the distance between the centroids of the bottom flange and the concrete deck. The flexural resistance provided by the flanges, M_{fl}, is then taken to be equal to the moment arm, A, times the bottom flange design force, P_{fy}.

The moment arm, A, for sections in negative flexure and for non-composite sections, is taken to be equal to the distance between the top and bottom flange centroids. The force to be used in calculating the flange moment is the smaller of P_{fy} for the top flange splice and P_{fy} for the bottom flange splice.

Should the Strength limit state moment, M_{uStr}, be larger than the moment resistance provided by the flanges, then the additional horizontal web splice force, H_w, is determined from Equation 6.45. For composite sections in positive flexure, A_w is equal to the distance between the mid-height of the web and the deck centroid. For non-composite sections and sections in negative flexure, A_w is equal to one-quarter of the web depth. The horizontal and vertical components of web splice force are combined in a vector fashion shown in Equation 6.46.

$$H_w = \frac{M_{uStr} - M_{fl}}{A_w} \tag{6.45}$$

$$R_{web} = \sqrt{V_r^2 + H_w^2} \tag{6.46}$$

For bolt shear resistance at the Strength limit state, the resistance factor, ϕ, is equal to 0.80, and Equation 6.47 or 6.48, as appropriate, provides the nominal resistance.

$$R_n = 0.56 A_b F_{ub} N_s \; (\text{threads excluded from shear planes}) \tag{6.47}$$

$$R_n = 0.45 A_b F_{ub} N_s \; (\text{threads not excluded from shear planes}) \tag{6.48}$$

The bolt tensile strength, F_{ub}, is 120 ksi for ASTM F3125, Grade A325 bolts, and 150 ksi for Grade A490 bolts. A_b is the bolt area corresponding to the nominal bolt diameter and N_s is the number of shear planes. If the distance between extreme bolts is greater than 38 inches, a penalty factor of 0.83 is to be applied to the bolt shear resistance, R_n. If filler plates 0.25 inches in thickness or greater are used, an additional penalty factor is to be applied to R_n. The filler plate penalty factor, R, is given by Equations 6.49 and 6.50. A_f is the area of the filler plate and A_p is the smaller of either (a) the flange area or (b) the sum of the splice plate areas.

$$R = \frac{1+\gamma}{1+2\gamma} \tag{6.49}$$

$$\gamma = \frac{A_f}{A_p} \tag{6.50}$$

Block shear in splice plates and flange plates must be checked. Equation 6.51 provides the nominal resistance. The design resistance, ϕR_n, incorporates a resistance factor, $\phi, = 0.80$, for block shear checks. U_{bs} is equal to 1.0 for splice plates in tension. R_p is 1.0 for splice plates since holes punched full size are not permitted in splices.

$$R_n = R_p \left(0.58 F_u A_{vn} + U_{bs} F_u A_{tn} \right) \le R_p \left(0.58 F_y A_{vg} + U_{bs} F_u A_{tn} \right) \tag{6.51}$$

$$R_p = \begin{cases} 0.90 \text{ for bolt holes punched full size} \\ 1.0 \text{ for bolt holes drilled full size} \\ 1.0 \text{ for bolt holes sub-punched and reamed to size} \end{cases} \tag{6.52}$$

Resistance factors for additional design checks include $\phi = 0.80$ net section rupture and $\phi = 0.95$ for gross section yielding of flange plates in tension. Nominal resistances for these limit states are given in Equations 6.53 and 6.54.

$$R_n = F_y A_g \text{ for yielding} \tag{6.53}$$

$$R_n = F_u A_n R_p U \text{ for rupture} \tag{6.54}$$

The shear lag factor, U, is equal to 1.0 for splice plates in tension.

For web splice plates in shear, Equations 6.55 for yielding and 6.56 for rupture are the applicable checks. The resistance factor for shear yielding is $\phi = 1.0$, and that for shear rupture is $\phi = 0.80$.

$$R_n = 0.58 F_y A_{gv} \text{ for yielding} \tag{6.55}$$

$$R_n = 0.58 F_u A_{vn} R_p \text{ for rupture} \tag{6.56}$$

Web bolts should be checked for slip under the Service II limit state factored shear at the splice, or the deck casting shear (multiplied by a typical load factor of 1.4), whichever is larger. If the flange bolts are not able to resist slip for the factored moment, then the web bolts will require additional slip checks. See the literature (Grubb et al., 2018) for details on such checks.

The combined areas of the flange splice plates must equal or exceed the area of the smaller flanges to which they are attached. Web plate areas must equal or exceed the web area.

Fatigue generally need not be checked for bolted field splice designs because the current version of the specification provisions is designed to preclude Fatigue failure in the splice. Refer to the AASHTO LRFD BDS commentary in Section 6.13.6.1.3 for further discussion.

Slip-critical provisions are to be met at the Service II limit state for the completed structure on the composite section. Slip-critical provisions are also to be satisfied on the non-composite section for loads during casting of the deck, with a typical load factor of 1.4 applied to the maximum deck casting moment. Slip resistance, R_n, is computed as shown in Equation 6.57.

$$R_n = K_s K_h N_s P_t \tag{6.57}$$

The required minimum bolt tension, P_t, the hole-related factor, K_h, and the surface-related factor, K_s, are given in Tables 6.3 through 6.5. N_s is the number of slip surfaces, typically equal to 2 for steel bridge girder field splices as long as the area on inner-to-outer splice plates is within 0.90 to 1.10 so that the flange splice force may be assumed to be equally distributed between inner and outer splice plates.

TABLE 6.3
Minimum Bolt Pre-tension, P_t (kips)

Bolt Diameter	Grade A325 Bolts	Grade A490 Bolts
⅝ inch	19	24
¾ inch	28	35
⅞ inch	39	49
1 inch	51	64
1 ⅛ inch	64	80
1 ¼ inch	81	102
1 ⅜ inch	97	121
1 ½ inch	118	148

TABLE 6.4
Slip Factor K_h

Standard holes	1.00
Over-sized and short-slotted holes	0.85
Long-slotted holes, slot perpendicular to force	0.70
Long-slotted holes, slot parallel to force	0.60

TABLE 6.5
Slip Factor K_s

Class A surface conditions	0.30
Class B surface conditions	0.50
Class C surface conditions	0.30
Class D surface conditions	0.45

Class A surface conditions refer to "unpainted clean mill scale, and blast-cleaned surfaces with Class A coatings" (AASHTO, 2020).

Class B surface conditions refer to "unpainted blast-cleaned surfaces to SSPC-SP 6 or better, and blast-cleaned surfaces with Class B coatings, or unsealed pure zinc or 85/15 zinc/aluminum thermal-sprayed coatings with a thickness less than or equal to 16 mils" (AASHTO, 2020).

Class C surface conditions refer to "hot-dip galvanized surfaces" (AASHTO, 2020).

Class D surface conditions refer to "blast-cleaned surfaces with Class D coatings" (AASHTO, 2020).

Bolt limitations include:

- center–center spacing $\geq 3d$
- center–center space on a free edge $\leq 4 + 4t \leq 7$ inches (t = thickness of thinner part)
- oversized and slotted holes not to be used
- no fewer than 2 rows of bolts on each side of the splice centerline
- minimum bolt size is ¾-inch diameter

Other checks to be made for splice bolts and splice plates include:

- bolt shear
- bearing on bolt holes
- slip of connected plates
- block shear
- gross section yielding
- net section fracture

6.7   STABILITY BRACING

The AASHTO LRFD BDS (AASHTO, 2020) does not outline stability bracing requirements in detail. Excellent guidance is available in Volume 13 of the Federal Highway Administration (FHWA) Steel Bridge Design Handbook (FHWA, 2015). Another valuable source of relevant information may be found in a recent National Cooperative Highway Research Program Report (National Academies of Sciences, Engineering, and Medicine, 2021). The material presented here is based largely on these documents along with American Institute of Steel Construction, AISC 360-16 (AISC, 2016). Lateral bracing of the compression flange in bridge girders is panel bracing, according to AISC 360-16, Appendix 6 terminology.

The required shear strength and shear stiffness of panel bracing are given by Equations 6.58 and 6.59, respectively (AISC Appendix 6, Equations A.6.5 and A.6.6a):

$$V_{br} = 0.01\left(\frac{M_u C_d}{h_o}\right) \qquad (6.58)$$

$$\beta_{br} \geq \frac{1}{\phi} \cdot \left(\frac{4M_u C_d}{L_{br} h_o} \right) \tag{6.59}$$

- ϕ = resistance factor = 0.75
- M_u = required LRFD flexural strength within the panel being considered, inch-kips
- C_d = 2.0 for the brace closest to the inflection point in a beam subject to double curvature bending
- C_d = 1.0 for all other cases
- L_{br} = unbraced length of the panel under consideration, inches
- h_o = distance between flange centroids, inches

The actual stiffness of the panel bracing may be estimated, assuming that the girder area is much larger than the brace area (typically a reasonable assumption), resulting in Equation 6.60.

$$\beta_{br} = \left(\frac{E}{S} \right) \left(\frac{4A_1 A_2 \cos^3 \theta}{4A_1 + A_2 \cos^3 \theta} \right) \tag{6.60}$$

Note that M_u/h_o is an estimate of the flange force. Conservatively, one could take M_u/h_o equal to F_y for the flange times the area of the flange.

A laterally braced panel is shown in Figure 6.4 in plan view.

Cross-frames and diaphragms are torsional bracing systems. Until AASHTO incorporates requirements for such systems into the AASHTO LFRD BDS, AISC Appendix 6 Equation A.6.9 gives the required strength of the torsional brace, repeated here in Equation 6.61a. A recent study (National Academies of Sciences, Engineering, and Medicine, 2021) concluded that this may provide unsatisfactory results, and Equation 6.61b is the recommended torsional brace strength from that study.

$$M_{br} = 0.02M_u \tag{6.61a}$$

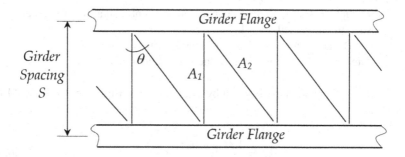

FIGURE 6.4 Lateral (panel) bracing of a bridge girder flange.

$$M_{br} = \frac{0.036M_uL}{nC_bL_b} \tag{6.61b}$$

Actual system stiffness, β_T, is composed of three springs in series and is given by Equation 6.62.

- brace stiffness, β_b
- web distortion stiffness, β_{sec}
- girder system stiffness, β_g

$$\frac{1}{\beta_T} = \frac{1}{\beta_b} + \frac{1}{\beta_{sec}} + \frac{1}{\beta_g} \tag{6.62}$$

Equation 6.63 provides the required system stiffness (from AISC Appendix 6, Equation A.6.11a). However, in accord with the National Academies report (National Academies of Sciences, Engineering, and Medicine, 2021), the coefficient 2.4 has been replaced with the coefficient 3.6 in the given equation, so there is a 50% difference between the equation shown in AISC and that given here.

$$\beta_T \geq \frac{1}{\phi} \cdot \frac{3.6L}{nEI_{yeff}} \cdot \left(\frac{M_u}{C_b}\right)^2 \tag{6.63}$$

- E = modulus of elasticity for steel, 29,000 ksi
- I_{yeff} = effective out-of-plane moment of inertia of the beam being braced, in^4

$$I_{yeff} = I_{yc} + \frac{t}{c}I_{yt} \tag{6.64}$$

- I_{yc} = moment of inertia of compression flange about y-axis, in^4
- I_{yt} = moment of inertia of tension flange about y-axis, in^4
- t = distance from neutral axis to extreme tension fiber, inches
- c = distance from neutral axis to extreme compression fiber, inches
- L = length of span being braced, inches
- ϕ = resistance factor = 0.75
- M_u = largest factored moment in unbraced segment, containing the brace, of the span being braced, inch-kips
- C_b = LTB (lateral torsional buckling) modification factor
- n = number of brace points within the span being braced

The web distortional stiffness is given by Equation 6.65 (Equation A.6.12 of AISC, Appendix 6).

$$\beta_{sec} = \frac{3.3E}{h_o}\left(\frac{1.5h_ot_w^3}{12} + \frac{t_{st}b_s^3}{12}\right) \tag{6.65}$$

- h_o=distance between flange centroids of the beam being braced, inches
- t_w=web thickness of the beam being braced, inches
- t_{st}=thickness of web stiffeners on the beam being braced, inches
- b_s=stiffener width for one-sided stiffeners; twice the individual stiffener width for pairs of stiffeners, inches

The stiffness of the torsional brace component of the system is provided by Equations 6.66 through 6.69 for various configurations (Yura, 2001).

$$\beta_b = \frac{6EI_b}{S} \text{ for diaphragms} \tag{6.66}$$

$$\beta_b = \frac{2ES^2h_b^2}{\dfrac{8L_c^3}{A_c} + \dfrac{S^3}{A_h}} \text{ for } K\text{-frame cross-frames} \tag{6.67}$$

$$\beta_b = \frac{ES^2h_b^2}{\dfrac{2L_c^3}{A_c} + \dfrac{S^3}{A_h}} \text{ for } Z\text{-frame cross-frames} \tag{6.68}$$

$$\beta_b = \frac{A_cES^2h_b^2}{L_c^3} \text{ for } X\text{-frame cross-frames} \tag{6.69}$$

The parameters used in the stiffness equations are as follows:

- S=girder spacing, inches
- A_h=area of horizontal cross-frame member, in^2
- A_c=area of diagonal cross-frame member, in^2
- L_c=length of diagonal cross-frame member, inches
- h_b=height of cross-frame, inches
- I_b=moment of inertia of diaphragm member, in^4

The equation for diaphragm stiffness (Equation 6.66), is based on the assumption that the diaphragm is located near the compression flange, forcing the diaphragm into double curvature bending under girder buckling. Should diaphragms be located near the tension flanges, as may be the case with railway through-girders and similar structures, then the equation must be modified to incorporate a coefficient of 2, rather than 6.

The girder system stiffness, β_g, is given by Equation 6.70.

$$\beta_g = \frac{24(n_g - 1)^2}{n_g} \cdot \frac{S^2EI_x}{L^3} \tag{6.70}$$

- I_x=strong axis moment of inertia for one girder of the girders being braced, in^4
- n_g=number of girders being braced
- S=spacing of the girders being braced, inches
- L=length of span being braced, inches

Torsional bracing may become ineffective in systems for which L_g/S becomes very large. The buckling capacity of a single girder of a twin-girder system in the global system buckling mode may be estimated from Equation 6.71 (Helwig and Yura, 2015).

$$M_{gs} = \frac{\pi^2 SE}{L_g^2} \sqrt{I_{yeff} I_x}$$ (6.71)

For a 3-girder system, Equation 6.71 may be used by replacing I_{yeff} with $1.5I_{yeff}$ and S with $2S$.

For a 4-girder system, Equation 6.71 may be used by replacing I_{yeff} with $2I_{yeff}$ and S with $3S$.

One method of increasing stability is through the use of combined lateral braces and torsional braces. End-panel bracing up to $b=0.20L$ has been shown to be effective.

6.8 SHEAR STUDS

Shear studs welded to the top flange of steel I-girders provide for composite action between the girder and the bridge deck.

The ratio of the height to the diameter of the shear must be no less than 4.0.

The center-to-center spacing of studs must not exceed 48 inches for members having a web depth greater than or equal to 24 inches. For members with a web depth less than 24 inches, the center-to-center spacing must not exceed 24 inches. The center-to-center spacing of studs is also limited to no less than six stud diameters.

Stud shear connectors must be no closer than 4.0 stud diameters center-to-center transverse to the longitudinal axis of the I-girder. The clear distance between the edge of the top flange and the edge of the nearest stud must not be less than 1 inch.

The clear depth of concrete cover over the top of studs must be at least 2 inches. Studs must extend at least 2 inches into the concrete deck.

Shear stud requirements include those for the Fatigue I limit state and for the Strength limit state.

Fatigue limit state requirements are summarized in Equations 6.72 through 6.76. Z_r is the resistance for a single stud. The required pitch (longitudinal spacing) of studs for the Fatigue limit state is p. The factor, w, is equal to 24 inches at end supports and 48 inches otherwise. F_{rc} is the range of cross-frame forces at the top flange for the Fatigue limit state. V_f is the vertical shear range due to Fatigue limit state loading. The number of shear connectors transverse to the girder centerline is n per row. The equation for F_{fat} is valid only for straight girders. Q_{ST} and I_{ST} are properties of the short-term composite section, calculated using a modular ratio, $n=E_S/E_C$

(not to be confused with the number of studs per row, n, appearing in Equation 6.73; context should reveal which n-parameter is under discussion).

$$Z_r = 5.5d^2 \tag{6.72}$$

$$p \le \frac{nZ_r}{V_{sr}} \tag{6.73}$$

$$V_{sr} = \sqrt{\left(V_{fat}\right)^2 + \left(F_{fat}\right)^2} \tag{6.74}$$

$$V_{fat} = \frac{V_f Q_{ST}}{I_{ST}} \tag{6.75}$$

$$F_{fat} = \frac{F_{rc}}{w} \tag{6.76}$$

Strength limit state requirements for shear studs are summarized in Equations 6.77 through 6.84. The minimum yield and tensile strengths of studs are specified in the AASHTO LRFD BDS, Section 6.4.4, to be $F_y = 50$ ksi and $F_u = 60$ ksi. For Strength limit state criteria, n is the number of total shear studs required between two locations under consideration. The cross-sectional area, A_{sc}, of a single shear stud is simply $\pi d^2/4$. For straight girders, both F_P and F_T may be taken to be equal to zero. For curved girders, the reader is referred to Section 6.10.10 of the AASHTO LRFD BDS.

Between points of zero and maximum positive moment, the required Strength limit state design shear, P_1, is based on P_P and F_P, and n gives the number of studs required between those points. Between points of maximum negative and maximum positive moment, the required Strength limit state design shear, P_2, is based on P_T and F_T, and n gives the number of studs required between those points.

$$Q_r = \phi_{sc} Q_n = 0.85 Q_n \tag{6.77}$$

$$Q_n = 0.5 A_{sc}\sqrt{f_c' E_c} \le A_{sc} F_u \tag{6.78}$$

$$n = \frac{P}{Q_r} \tag{6.79}$$

$$P_1 = \sqrt{P_P^2 + F_P^2} \tag{6.80}$$

$$P_2 = \sqrt{P_T^2 + F_T^2} \tag{6.81}$$

$$P_P = \text{Min} \begin{cases} 0.85 f_c' b_s t_s \\ Dt_w F_{yw} + F_{yt} b_{ft} t_{ft} + F_{yc} b_{fc} t_{fc} \end{cases} \tag{6.82}$$

$$P_T = P_P + P_n \tag{6.83}$$

$$P_n = \text{Min} \begin{cases} 0.45 f'_c b_s t_s \\ Dt_w F_{yw} + F_{yt} b_{ft} t_{ft} + F_{yc} b_{fc} t_{fc} \end{cases} \tag{6.84}$$

6.9   PLASTIC MOMENT COMPUTATIONS

Plastic moment computations are based on yielding of the entire cross section, not simply the most remote fibers of the cross section. Concrete stress is set to be equal to $0.85 f'_c$ and all steel elements, including any reinforcement included in the calculation, are stressed to their respective yield stress, f_y. Under the full yielding cross-sectional assumption, the location within the cross-section depth at which the forces above exactly equal the forces below is the plastic neutral axis (PNA).

For non-composite sections, the calculations are based solely on the girder properties and are rather straightforward.

For composite sections in negative flexure, given that the plastic moment is a Strength or Extreme Event limit state, the concrete deck is typically not included, but the deck reinforcement may be included in cases where adequate shear studs are provided in the negative moment region.

For composite sections in positive flexure, it is not uncommon to discount any reinforcement in the concrete deck since such reinforcement is typically reduced from that in negative moment regions so that a smaller influence on the deck reinforcement occurs. Nonetheless, deck reinforcement may be included, provided the construction documents clearly outline the required reinforcing details. Certainly, the concrete deck is included in plastic moment calculations for girders designed to act composite with the deck through adequate shear stud provision.

6.10   SOLVED PROBLEMS

Problem 6.1

Using the NSBA LRFD-Simon software, create a continuous beam model of the steel superstructure for an interior girder of the Project Bridge. Use a welded plate I-girder with constant properties over the entire bridge length. In practice, design optimization would include flange thickness and/or width transitions to avoid material waste. For this problem use the following properties and parameters:

- $b_f = 16$ inches, $t_f = 1.25$ inches, top and bottom flanges
- $D = 46$ inches, $t_w = 0.50$ inches, web
- $F_y = 50$ ksi, $F_u = 70$ ksi, flanges and web
- $f'_c = 4$ ksi, $n = 8$, $3n = 24$, composite deck
- $A_s = 9.2$ in^2, deck reinforcement in negative moment regions
- Top of the web to the bottom of the deck $= 3.25$ inches
- Assume metal deck forms add a weight equal to 1 inch of concrete
- Use weathering steel for all plates
- Pour #1 is from the ends of the bridge to the field splice
- Pour #2 is from the field splice to the pier

- Parapet weight=0.400 klf per parapet
- Girder spacing S=9 ft 3 inches
- Try for an unstiffened web design

Problem 6.2

Using the results reported by Simon for Problem 6.1, verify the section properties for the girder, the girder with rebar, the short-term composite properties, and the long-term composite properties.

Problem 6.3

Using the reported moments from Simon for Problem 6.1, check the stresses reported for the Strength I limit state at the points of maximum positive moment and maximum negative moment.

Problem 6.4

Using the reported shear results from Problem 6.1, check the reported shear resistances at the end of the bridge and at the pier.

Problem 6.5

Using results from the Simon model in Problem 6.1, verify the Fatigue I limit state reported ratios for each of the following at the maximum positive moment section:

a) base metal, bottom flange, Category B
b) web-to-flange weld, bottom flange, Category B
c) stiffener-to-flange weld, bottom flange, Category C'

Problem 6.6

Use three 0.75-inch diameter studs per row on the top 16-inch×1.25-inch flange for the Project Bridge steel girder. Using fatigue shear range and section property results from the Simon model from Problem 6.1, determine:

- the shear stud pitch requirements for the Fatigue I limit state
- the number of shear studs between points of zero and maximum positive moment for the Strength limit state
- the number of shear studs between the points of maximum positive and maximum negative moment for the Strength limit state.

Problem 6.7

For the steel girder of the Project Bridge, use Simon results from Problem 6.1 to verify the moment resistance at the section 36.59 feet from the abutment end. The following parameters are given for the plastic condition; other necessary parameters are to be found in the Simon results.

- M_P=9,025 ft-kips
- y_P=48.49 inches, bottom of girder to plastic neutral axis (PNA)

Problem 6.8

A welded steel plate I-girder consists of 75-inch×0.50-inch web with 20-inch×1-inch flanges, top and bottom. All I-girder plates are Grade 70W weathering steel. Assume that the required design shear is equal to the design shear resistance and the design two-sided transverse stiffeners using Grade 50W steel. Check two-sided bearing stiffeners, 9.5 inches×0.875 inches, Grade 50W, with 1-inch clips, for a Strength limit state reaction at an interior Pier of 900 kips.

Deck, $(9.25/12)(.15)(34.42)/4 = 0.995$ KLF
(per girder)

Haunch, $(2/12)(16/12)(.15) = 0.033$ KLF
(per girder)

Girder area $= 46 \times \frac{1}{2} + 2 \times 16 \times 1.25$
$= 63.0$ in^2

$W = \frac{63}{144} \times 490 = 214$ plf

The girder weight is automatically included in SIMON. However, to account for bracing & other misc. items, add 7.5% of the girder weight as part of the non-composite dead load (DC1).

DC1 $= 0.995 + 0.033 + 0.075 \times 0.214$

$\Rightarrow$ DC1 $= 1.044$ KLF
(per girder)

DC2 $= 2 \times 0.400/4 = 0.200$ KLF
(per girder)

DW $= 32 \times .035/4 = 0.280$ KLF
(per girder)

$d_e = 3'\text{-}4'' - 1'2\frac{1}{2}'' = 2.125'$

"CEE4380 - Steel Project - Interior. dat"

PROBLEM 6.1	TEH	2/5

Comments, line 1	CEE 4380
Comments, line 2	Bridge Design
Comments, line 3	Tim Huff
Beam type	I-Girder ∨
Number of spans	2
Number of girders	4
Number of traffic lanes	2
Run option	LRFD Analysis ∨
Redesign performance ratio	0.9
Maximum performance ratio	1.02
Minimum flange thickness	0.75 in
Maximum plate thickness	4.00 in
Distance from slab bottom to cg of reinforcement	4.25 in
Distance from slab bottom to web top	3.25 in
Average daily truck traffic, single lane	5000
Fatigue service life	75 years

Distribution factor definition Program Defined ∨

Computed Distribution Factors

Girder skew	0	degrees
Girder spacing	9.25	ft
Distance from web to curb, de	2.125	ft
Girder location	Interior ∨	

PROBLEM 6.1	TEH	3/5

Modular ratio, n   `8`

Slab compressive strength   `4000`   psi

Reinforcement yield strength   `60`   ksi

Longitudinal stiffener yield strength   `50`   ksi

Transverse and bearing stiffener yield strength   `50`   ksi

Concrete type   Normal weight concrete   ⌄

Steel surface condition   Weathering steel   ⌄

Connection plate type   Welded connection plates   ⌄

Slab meet 6.10.1.7 criteria   Yes   ⌄

Uniform Dead Loads

Composite   `200`   lb/ft

Utility   `        `   lb/ft

Future wearing surface   `280`   lb/ft

Design vehicle option   HL93/User Defined Design Vehicle (envelope)   ⌄

Live load deflection factor   `800`

Pedestrian live load   `        `   lb/ft

Dynamic Load Allowance

Design vehicle   `1.33`

Fatigue vehicle   `1.15`

PROBLEM 6.1	TEH	4/5

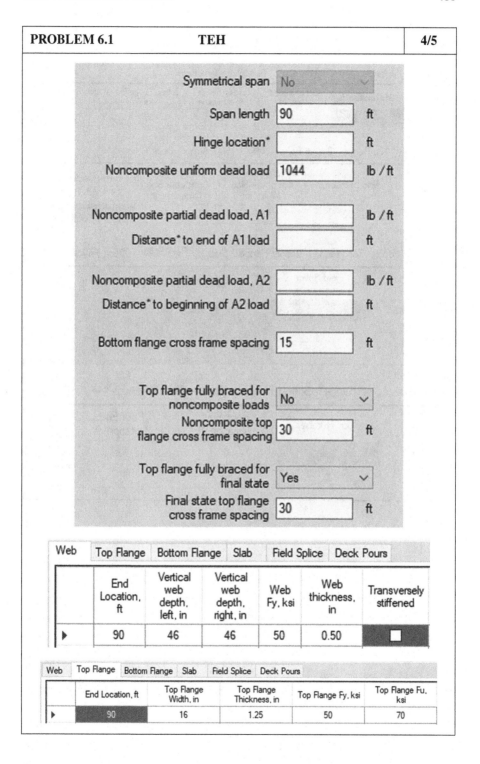

Symmetrical span   No

Span length   90   ft

Hinge location*   ft

Noncomposite uniform dead load   1044   lb / ft

Noncomposite partial dead load, A1   lb / ft

Distance* to end of A1 load   ft

Noncomposite partial dead load, A2   lb / ft

Distance* to beginning of A2 load   ft

Bottom flange cross frame spacing   15   ft

Top flange fully braced for noncomposite loads   No

Noncomposite top flange cross frame spacing   30   ft

Top flange fully braced for final state   Yes

Final state top flange cross frame spacing   30   ft

Web | Top Flange | Bottom Flange | Slab | Field Splice | Deck Pours

	End Location, ft	Vertical web depth, left, in	Vertical web depth, right, in	Web Fy, ksi	Web thickness, in	Transversely stiffened
▶	90	46	46	50	0.50	☐

Web | **Top Flange** | Bottom Flange | Slab | Field Splice | Deck Pours

	End Location, ft	Top Flange Width, in	Top Flange Thickness, in	Top Flange Fy, ksi	Top Flange Fu, ksi
▶	90	16	1.25	50	70

| PROBLEM 6.1 | TEH | 5/5 |

Web | Top Flange | Bottom Flange | Slab | Field Splice | Deck Pours

	End Location, ft	Bottom Flange Width, in	Bottom Flange Thickness, in	Bottom Flange Fy, ksi	Bottom Flange Fu, ksi
▶	90	16	1.25	50	70

Web | Top Flange | Bottom Flange | Slab | Field Splice | Deck Pours

	End Location, ft	Effective Composite Slab Width, in	Effective Composite Slab Thickness, in	Reinforcement Area, A's, in^2
▶	90	111	8.25	9.20

Web | Top Flange | Bottom Flange | Slab | Field Splice | Deck Pours

	Field Splice Location, ft
▶	63

Web | Top Flange | Bottom Flange | Slab | Field Splice | Deck Pours

	Pour Number	Pour Start Location, ft	Pour End Location, ft
▶	1	0	63
	2	63	90

PROBLEM 6.2	TEH	1/4

Girder

$$A = 46 \times \tfrac{1}{2} + 2 \times 16 \times 1\tfrac{1}{4} = 63.0 \, in^2$$

From symmetry, $\bar{y} = (46 + 2 \times 1\tfrac{1}{4})/2$

$$\Rightarrow \bar{y} = 24.25''$$

$$I = .5(46)^3/12 + 2(16)(1.25)^3/12$$
$$+ 2(16 \times 1.25)(23.625)^2$$

$$\Rightarrow I = 26,387 \, in^4$$

$$S_{top} = S_{bott} = 26,387/24.25$$

$$\Rightarrow S_{top} = S_{bott} = 1,088 \, in^3$$

	Hand Calc	Simon
A	63.0	
I	26,387	26,381
S_{top}	1,088	1,088
S_{bott}	1,088	1,088
$\bar{y}$	24.25	24.25

PROBLEM 6.2	TEH	2/4

Girder + Rebar

$$A = 9.2 + 63.0 = 72.2 \, in^2$$

$$\bar{y} = \frac{9.2(54.75) + 63.0(24.25)}{72.2}$$

$$\Rightarrow \bar{y} = 28.14 \, in$$

$$I = 26,381 + 63(28.14 - 24.25)^2$$
$$+ 9.2(54.75 - 28.14)^2$$

$$\Rightarrow I = 33,849 \, in^4$$

$$S_{top} = \frac{33,849}{48.5 - 28.14} = 1,662 \, in^3$$

$$S_{bott} = \frac{33,849}{28.14} = 1,203 \, in^3$$

	Hand Calc	Simon
A	72.2	
I	33,849	33,849
S_{top}	1,662	1,662
S_{bott}	1,203	1,203
$\bar{y}$	28.14	28.14

PROBLEM 6.2	TEH	3/4

Short-term Composite $(n=8)$

$b' = 111/8 = 13.875''$

$A = 13.875(8.25) + 63.0 = 177.47 \text{ in}^2$

$\bar{y} = \dfrac{13.875(8.25)(54.625) + 63(24.25)}{177.47}$

$\Rightarrow \bar{y} = 43.84''$

$I = 13.875\dfrac{(8.25)^3}{12} + 13.875(8.25)(10.78)^2$

$\quad + 26,381 + 63.0(19.59)^2$

$\Rightarrow I = 64,510 \text{ in}^4$

$S_{top} = \dfrac{64,510}{48.5 - 43.84} = 13,843 \text{ in}^3$

$S_{bott} = \dfrac{64,510}{43.84} = 1,471 \text{ in}^3$

	Hand Calc	Simon
A	177.5	
I	64,510	64,648
S_{top}	13,843	14,261
S_{bott}	1,471	1,470
$\bar{y}$	43.84	43.97

PROBLEM 6.2	TEH	4/4

Long-term Composite $(3n = 24)$

$b' = 111/24 = 4.625''$

$A = 4.625(8.25) + 63 = 38.16 + 63 = 101.2$

$$\bar{y} = \frac{38.16(54.625) + 63(24.25)}{101.2}$$

$$\Rightarrow \bar{y} = 35.71''$$

$I = 4.625(8.25)^3/12 + 38.16(18.915)^2$

$\quad + 26,381 + 63(11.46)^2$

$$\Rightarrow I = 48,523 \text{ in}^4$$

$S_{top} = \dfrac{48,523}{48.5 - 35.71} = 3,794 \text{ in}^3$

$S_{bott} = \dfrac{48,523}{35.71} = 1,359 \text{ in}^3$

	Hand Calc	Simon
A	101.2	
I	48,523	48,773
S_{top}	3,794	3,867
S_{bott}	1,359	1,359
$\bar{y}$	35.71	35.89

PROBLEM 6.3	TEH	1/2

M_{MAX}^{+} is at $X = 36.59$ ft from end

$$M_{DC1} = 121.5 + 591.4 = 712.9 \text{ ft·k}$$

$$M_{DC2} = 113.3 \text{ ft·k}$$

$$M_{DW} = 158.6 \text{ ft·k}$$

$$M_{LL+IM} = 1,369.5 \text{ ft·k}$$

DC1 :   Girder alone

$$S_{top} = S_{bott} = 1,088 \text{ in}^3$$

$$f_{top} = \frac{-1.25(712.9)}{1,088} \times 12 = -9.83 \text{ ksi}$$

$$f_{bott} = +9.83 \text{ ksi} \quad ("+" = \text{tension})$$

DC2 + DW :   Long-term Composite

$$S_{top} = 3,867 \text{ in}^3$$

$$S_{bott} = 1,359 \text{ in}^3$$

$$M_{STR-I} = 1.25(113.3) + 1.5(158.6)$$
$$= 379.5 \text{ ft·k}$$

$$f_{top} = \frac{-379.5(12)}{3,867} = -1.18 \text{ ksi}$$

$$f_{bott} = \frac{+379.5(12)}{1,359} = +3.35 \text{ ksi}$$

PROBLEM 6.3	TEH	2/2

LL+IM : Short-term Composite

$$S_{top} = 14,261 \text{ in}^3$$

$$S_{bott} = 1,470 \text{ in}^3$$

$$f_{top} = \frac{-1.75(1,369.5)}{14,261} \times 12 = -2.02 \text{ ksi}$$

$$f_{bott} = \frac{+1.75(1369.5)}{1,470} \times 12 = 19.56 \text{ ksi}$$

$$f_{top} = -9.83 + (-1.18) + (-2.02) = -13.03 \text{ ksi}$$

$$f_{bott} = +9.83 + 3.35 + 19.56 = +32.74 \text{ ksi}$$

	Hand Calc	Simon
f_{top}	-13.03 ksi	-13.02 ksi
f_{bottom}	+32.74 ksi	+32.74 ksi

("+" stress = tensile stress)

PROBLEM 6.4	TEH	1/2

$\underline{\text{Bridge End}}$, $x = 0.0$ ft

$V_{DC1} = 7.2 + 35.3 = 42.5$ kips

$V_{DC2} = 6.8$ kips

$V_{DW} = 9.5$ kips

$V_{LL+IM} = 97.9$ kips

$V_u = 1.25(42.5 + 6.8) + 1.50(9.5)$
$+ 1.75(97.9)$

$\Rightarrow \underline{V_u = 247.2 \text{ kips}}$

$k = 5.0 \qquad \sqrt{\dfrac{Ek}{F_{yw}}} = \sqrt{\dfrac{29,000 \times 5}{50}}$

$= 53.8$

$D/t_w = 46/.5 = 92.0$

$1.4 \times 53.8 = 75.4$

$92.0 > 75.4 \Rightarrow C = 1.57 \left(\dfrac{53.8}{92.0}\right)^2$

$\Rightarrow C = 0.537$

$\phi V_n = 1.0 \times 0.537 \times 0.58 \times 50 \times 46 \times .5$

$\Rightarrow \phi V_n = 358.2$ kips

$V_u / \phi V_n = 247.2 / 358.2 = 0.690$

(Simon reports 0.689)

PROBLEM 6.4	TEH	2/2

At Pier, $x = 90.0$ ft

$V_{DC1} = 12.1 + 58.7 = 70.8$ kips

$V_{DC2} = 11.2$ kips

$V_{DW} = 15.7$ kips

$V_{LL+IM} = 112.9$ kips

$V_u = 1.25(70.8 + 11.2) + 1.50(15.7)$
$\quad + 1.75(112.9)$

$\Rightarrow V_u = 323.6$ kips

$\dfrac{V_u}{\phi V_n} = \dfrac{323.6}{358.2} = 0.903$

(Simon reports 0.902)

PROBLEM 6.5	TEH	1/2

Bottom of Bottom Flange,

$$S_{bott} = 1,470 \ in^3$$
(short-term composite)

$$\bar{y} = 43.97''$$

$\Rightarrow$ Top of bottom flange, S'_{bott}:

$$S'_{bott} = 1,470 \left(\frac{43.97}{43.97 - 1.25} \right)$$

$$= 1,513 \ in^3$$

a) Base Metal, Category B, $\gamma = 1.75$

$$\Delta F_n = \Delta F_{TH} = 16 \ ksi$$

$$M_{fat} = 422.0'-k, \ Max$$
$$\quad\quad -101.1^{'-k}, \ Min$$

$$M_{range} = 422.0 - (-101.1)$$

$$= 523.1 \ ft \cdot k$$

$$\Delta f = 523.1 \times 12 / 1,470 = 4.27 ksi$$

$$\gamma \Delta f = 1.75 \times 4.27 = 7.47 \ ksi$$

$$\frac{\gamma \Delta f}{\Delta F_n} = \frac{7.47}{16} = 0.467$$

PROBLEM 6.5	TEH	2/2

b) Flange-Web weld, Category B

$$\Delta F_n = \Delta E_{TH} = 16 \text{ ksi}$$

$$\Delta f = 523.1 \times 12 / 1,513 = 4.15 \text{ ksi}$$

$$\gamma \Delta f = 1.75 \times 4.15 = 7.26 \text{ ksi}$$

$$\frac{\gamma \Delta f}{\Delta F_n} = \frac{7.26}{16} = 0.454$$

c) Stiffener-to-Flange weld
Category C', $\Delta F_n = \Delta E_{TH} = 12 \text{ ksi}$

$$\frac{\gamma \Delta f}{\Delta F_n} = \frac{7.26}{12} = 0.605$$

	Hand Calc	Simon
a)	0.467	0.467
b)	0.454	0.454
c)	0.605	0.605

PROBLEM 6.6	TEH	1/5

$$Z_r = 5.5 \, (.75)^2 = 3.09 \, ^k/stud$$

From Simon:

$$V_r = V_{range} \times 1.75 \; (Fatigue \, I)$$

X	Vrange	$V_r = V_{range} \times 1.75$ (Fatigue I)
0	42.1^k	73.7^k
9	36.2^k	63.4^k
18	32.3^k	56.5^k
27	30.3^k	53.0^k
36	28.7^k	50.2^k
45	29.4^k	51.5^k
54	31.4^k	55.0^k
63	34.3^k	60.0^k
72	37.1^k	64.9^k
81	40.0^k	70.0^k
90	42.8^k	74.9^k

Short-term Composite:

$$I = 64,648 \, in^4$$

$$\bar{y} = 43.97''$$

$$46 + 2(1.25) + 2 + 8.25/2$$
$$-43.97 = 10.66 \; inches$$

43.97

PROBLEM 6.6	TEH	2/5

$$\overline{y} = 10.66'' \text{ at deck/girder int.}$$

$$A = (11\frac{1}{8})(8.25) = 114.5 \text{ in}^2$$

$$\overline{y} A = 1,220 \text{ in}^3$$

3 studs per row, $n = 3$

$$P = \frac{3 \times 3.09}{V_{fat}}$$

$$V_{fat} = V_r (10.66 \times 114.5)/64,648$$

$$= V_r / 52.93$$

$$P = \frac{3 \times 3.09 \times 52.93}{V_r}$$

$$P = 6'' \Rightarrow V_r \le 81.8 \text{ kips}$$

$$P = 9'' \Rightarrow V_r \le 54.5 \text{ kips}$$

6" pitch ok at abutment end.
9" pitch ok 27' from abut. end.
Back to 6" pitch 49.5' from end.

For negative contraflexure
regions, use girder + rebar.

PROBLEM 6.6	TEH	3/5

Girder + rebar :

$$I = 33,849 \ in^4$$

$$y_b = 28.14''$$

$$A = 9.2 \ in^2 \ , \ reinforcement \ in \ deck \ per \ beam$$

26.5''

28.14''

$$Q = \bar{y} A$$

$$= 26.5 \times 9.2$$

$$= 244 \ in^3$$

$$V_{fat} = V_r \ (244)/33,849$$

$$= V_r/139$$

$$p = \frac{3 \times 3.09 \times 139}{V_r}$$

$$p = 18'' \ , \ V_r \leqslant 71.6 \ kips$$

$$p = 15'' \ , \ V_r \leqslant 85.9 \ kips$$

use $p = 15''$ at Pier

PROBLEM 6.6	TEH	4/5

Strength Limit State

$$A_{sc} = \pi (.75)^2 / 4 = 0.442 \, in^2$$

$$Q_n = .5 (0.442) \sqrt{4 \times 3640} = 26.6 \, k$$

$$\leq 0.442 \times 60 = 26.5^k$$

$$\Rightarrow \text{Take } Q_n = 26.5^k/stud$$

$$Q_r = \phi Q_n = .85 \times 26.5$$

$$= 22.5 \, k/stud$$

Between abutment and ⅊ Max. Positive
Moment at 36.59 ft from end:

$$P_p = Min \begin{cases} .85 \times 4 \times 111 \times 8.25 = 3,114 \, kips \\ 63 \times 50 = 3,150 \, kips \end{cases}$$

$$\Rightarrow P_p = 3,114 \, kips$$

$$n = 3,114 / 22.5 = 138.4 \, studs$$

$$\div 3 = 46.1 \, rows$$

Between Max Positive Moment at
36.59 ft from end and Pier at
90 ft from end:

$$P_n = Min \begin{cases} .45 \times 4 \times 111 \times 8.25 = 1,648^k \\ 63 \times 50 = 3,150 \, k \end{cases}$$

PROBLEM 6.6	TEH	5/5

$$P_T = P_p + P_n = 3,114 + 1,648$$

$$= 4,762 \text{ KIPS}$$

$$n \geqslant 4,762 / 22.5$$

$$= 212 \text{ studs}$$

$$\div 3 \text{ studs /row}$$

$$= 70.5 \text{ rows}$$

Detail stud spacing to satisfy
Fatigue I Limit State; also
ensure that the resulting
arrangement provides at least:

47 rows, 3 studs /row
from 0 ft to 36.59 ft

71 rows, 3 studs /row
from 36.59 ft to 90 ft

PROBLEM 6.7	TEH	1/1

$D_t = 46 + 2 \times 1.25 + 2 + 8.25 = 58.75''$

$D_p = 58.75 - 48.49 = 10.26''$

$D_p/D_t = \dfrac{10.26}{58.75} = 0.175 < .42$

$\rightarrow$ Ductility ok

$Ratio = 0.175 / 0.42 = 0.416$

$M_n = M_p (1.07 - 0.7 \times 0.175)$

$\quad = 9,025 (0.948)$

$\quad = 8,551 \; ft \cdot kips$

$M_n \leq 1.3 \, M_y \qquad M_y = 5,782 \, ft \cdot k$
$\qquad\qquad\qquad\qquad (Simon \; output)$

$M_n \leq 1.3 \times 5,782 = 7,517 \, ft \cdot k$
$\qquad\qquad\qquad < 8,551$

$\Rightarrow \phi M_n = 1.0 \times 7,517$
$\qquad\qquad = 7,517 \; ft \cdot k$

$M_u = 3,667 \; , \; Strength \; I \; , \; from \; Simon$

$M_u/\phi M_n = 3,667/7,517 = 0.488$

	Hand Calc	Simon
Ductility Ratio	0.416	0.416
$M_u / \phi M_n$	0.488	0.488

PROBLEM 6.8	TEH	1/3

$b_t \geqslant 2 + 75/30 = 4.50''$

$b_t \geqslant b_f/4 = 20/4 = 5.00''$

$t_p \geqslant b_t/16$

Web $F_y = 70$ ksi

stiffener $F_y = 50$ ksi

Try $b_t = 5''$ $t_p = 5/16''$

$b_t/t_p = 16$

$F_{crs} = 0.31 \times \dfrac{29,000}{16^2} = 35.1$ ksi

$P_t = \dfrac{70}{35.1} = 1.993$

$I_t \geqslant \dfrac{25^4}{40}(1.993^{1.3})\left(\dfrac{70}{29,000}\right)^{1.5}$

$\Rightarrow I_t \geqslant 230$ in^4

$I_t = 5/16(2\times5 + 0.5)^3/12 - 5/16(.5)^3/12$

$\Rightarrow I_t = 30.1$ in^4 < 230, No Good

Try $7\frac{1}{2}'' \times 9/16''$ stiffener

$b_t/t_p = 7.5/.5625 = 13.33 < 16$
ok

$b_t > 5.00'' = (b_f)_{min}$, ok

$b_t < \frac{1}{2}(20 - .5) = 9.75'' = (b_f)_{max}$
ok

PROBLEM 6.8	TEH	2/3

$$F_{crs} = .31 \times \frac{29,000}{13.33^2} = 50.6 \text{ ksi} > F_{ys}$$

$$\Rightarrow \overline{F}_{crs} = 50 \text{ ksi}$$

$$P_T = \text{Max} \begin{cases} 1.0 \\ 70/50 = 1.40 \end{cases} \leftarrow$$

$$(I_t)_{req'd} = \frac{75^4 (1.4)^{1.3}}{40} \left(\frac{70}{29,000}\right)^{1.5}$$

$$= 145.3 \text{ in}^4$$

$$(I_t)_{actual} = \frac{9}{16} \left\{ (2 \times 7.5 + .5)^3 - .5^3 \right\} / 12$$

$$= 174.5 \text{ in}^4 > 145.3$$

$$\Rightarrow OK$$

Use $7\frac{1}{2}'' \times {}^9/_{16}''$
50W Stiffener Pairs
(two-sided)

Bearing Stiffener:

$$b_t/t_p = 9.5/0.875 = 10.86$$

$$b_t/t_p \leq 0.48 \sqrt{\frac{29,000}{50}} = 11.56$$

$$10.86 < 11.56 , \underline{OK}$$

$$A_{pn} = 2(.875)(9.5-1)$$

$$\Rightarrow A_{pn} = 14.875 \text{ in}^2$$

PROBLEM 6.8	TEH	3/3

"Column" $A = 2(9.5)(.875) + 18(.5)(.5)$

$$= 21.125 \text{ in}^2$$

$I = .875(2 \times 9.5 + .5)^3 / 12$

$$= 540.67 \text{ in}^4$$

$r = \sqrt{\dfrac{540.67}{21.125}} = 5.06 \text{ in}.$

$KL/r = .75 \times 75 / 5.065 = 11.12$

$P_e = \pi^2 (29,000)(21.125) / (11.12)^2$

$\Rightarrow P_e = 48,908 \text{ kips}$

$P_o = 50 \times 21.125 = 1,056 \text{ kips}$

$P_e / P_o = 46.3 > 0.44$

$\Rightarrow P_n = 0.658^{(1/46.3)} (1,056)$

$$= 1,046 \text{ kips}$$

$\phi P_n = 0.95 \times 1046 = 994^k$

$$> 900^k, \underline{\underline{ok}}$$

$\phi (R_{sb})_n = 1.0 \times 1.4 \times 14.875 \times 50$

$\Rightarrow \phi (R_{sb})_n = 1,041^k > 900^k$

$$\Rightarrow \underline{\underline{ok}}$$

6.11 EXERCISES

E6.1

A welded plate girder made from Grade 50W weathering steel has a
52-inch × 1/2-inch web. Compute the design shear resistance, ϕV_n, for each
of the following conditions:

a)   an unstiffened end panel
b)   a stiffened end panel with stiffener spacing, $d_o = 1.5D$
c)   a stiffened interior panel with stiffener spacing, $d_o = 1.5D$

E6.2

Using the reported moments from the Simon model in Problem 6.1, verify
the reported stresses in the flanges over the Pier at the Strength I limit state
for the Project Bridge.

E6.3

For the Project Bridge, a field splice using 16-inch × 3/4-inch outer plate
at each flange and two 6.5-inch × 1-inch inner plates at each flange is proposed. The proposed web splice plates are 42 inch × 3/8 inch. Use Grade
50W splice plates and assume four bolts (A325, threads included, and
Class B surface) across the width of the flange. The girder consists of
16-inch × 1.25-inch flanges and a 46-inch × 0.50-inch web, all made from
Grade 50W plate. Assume that the web is unstiffened at the splice point.

- Check the flange splice plate areas for total area requirements and for
  equal distribution of design force requirements.
- Check the web splice plate areas for total area requirements.
- Determine the flange splice design force, P_{fy}.
- Determine the Strength I limit state maximum and minimum moments
  at the splice point, using Simon results from Problem 6.1.
- Determine the moment resistance provided by the flanges.
- Determine the web design force, R_{web}.
- Determine the number of flange bolts and the number of web bolts
  based strictly on bolt shear.
- Enter the data onto the NSBA-Splice spreadsheet to perform a complete
  check of the proposed design.

E6.4

A welded plate I-girder is used as the design basis for a three-span
bridge. Span lengths are 234 ft, 300 ft, and 234 ft. At a particular point
on the span, the girder cross section consists of 103-inch × 0.75-inch web,
18-inch × 1-inch top flange, 20-inch × 1.0625-inch bottom flange, and 8-inch
concrete deck. The distance from the top of the web to the bottom of the
deck is 5.5 inches. The unfactored moments, per girder, are summarized as
follows:

- $M_{DC1} = 4,823$ ft-kips
- $M_{DC2} = 448$ ft-kips
- $M_{DW} = 1,137$ ft-kips
- $M_{LL+IM} = 7,945$ ft-kips, $-2,720$ ft-kips
- $M_{Fatigue} = 2,349$ ft-kips, -713 ft-kips (both values include impact)

TABLE E6.4
Exercise E6.4

Property	Girder	Short-term Composite	Long-term Composite
I, in^4	174,249	431,434	312,546
y_b, inches	51.11	86.55	70.20
S_{top}, in^3	3,229	23,309	8,965
S_{bott}, in^3	3,410	4,985	4,452

The elastic section properties of the girder at various stages are given in Table E6.4. Yield moment, plastic moment, and distance from the girder bottom to the PNA are:

- $M_y = 17,713$ ft-kips
- $M_P = 30,599$ ft-kips
- $y_P = 104.2$ inches

Determine each of the following:

- the stresses at the bottom and the top of the girder at the Strength limit state.
- the fatigue stress range and acceptable category for a connection stiffener to bottom flange weld for the Fatigue I (infinite life) limit state.
- the required moment resistance and design moment resistance for the compact section in positive flexure at the Strength limit state.

E6.5

A three-span, steel I-girder bridge consists of the cross section shown in Figure E6.5 with the following parameters for an interior girder:

- $b_{eff} = 126$ inches
- $t_s = 8$ inches
- $x = 5.5$ inches, from the top of the web to the bottom of the deck

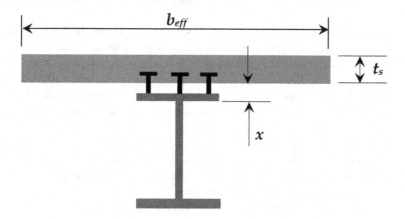

FIGURE E6.5   Exercise E6.5.

TABLE E6.5

Exercise E6.5

Plate	Left of Splice Centerline	Right of Splice Centerline
Top flange (inches)	22×1.5625	19×0.8750
Web (inches)	109×0.7500	109×0.750
Bottom flange (inches)	24×2.2500	19×0.9375

Span lengths are 234 ft, 300 ft, and 234 ft. The cross section to the left and the right of the centerline of a field splice 75 feet into span 2 are tabulated in Table E6.5.

The concrete for the deck has a 28-day minimum strength of 5 ksi. All plates are Grade 50W steel. Unfactored moments at the splice location are as follows:

- $M_{DC1} = -741$ ft-kips, non-composite
- $M_{DC2} = -7$ ft-kips, long-term composite
- $M_{DW} = -13$ ft-kips, long-term composite
- $M_{LL+IM} = 5,252$ ft-kips (maximum) and $-4,392$ ft-kips (minimum), short-term composite
- $M_{DeckPour} = -3,810$ ft-kips

Include effects from any required filler plates, and (a) check the proposed splice plates for required area and equal load distribution and (b) set the number of bolts for shear. Enter the data onto the NSBA Splice spreadsheet and perform a full check on the proposed splice design. The proposed splice design is summarized below.

- Outer splice plates for flange $= 19 \times 9/16$, top and bottom
- Inner splice plates for flange $= 8.125 \times 5/8$, top and bottom
- Web splice plates $= 102.75 \times 3/8$, stiffener spacing $d_o = 12$ ft
- 7/8-inch A325-N bolts, 4 rows for flanges and 2 rows each side of the web

E6.6

The framing plan for the center 300-ft span of a three-span bridge is shown in Figure E6.6a. A typical intermediate cross-frame between any two adjacent girders is shown in Figure E6.6b. The moments during construction are estimated to be 3,729 ft-kips from component dead load and 4,100 ft-kips from estimated construction live loads prior to deck curing. The chord and diagonal members of the cross-frames each have an area equal to 5.00 in². Determine the web distortional stiffness, β_{sec}, the girder system stiffness, β_g, and the cross-frame stiffness, β_b. Compute the actual total system stiffness and the required system stiffness, β_T. Determine the K-frame chord and diagonal design forces. Additional data are summarized below.

- Web plate $= 109$ inches $\times 0.75$ inches
- Flange plate $= 30$ inches $\times 1.25$ inches, top and bottom
- Stiffeners $= 8$ inches $\times 0.75$ inches, two sided

(a) c.l. Pier   c.l. Pier

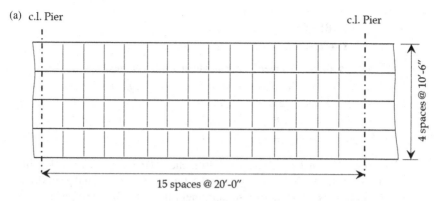

Partial Framing Plan

(b)

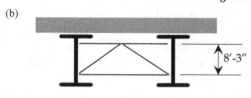

Partial Cross-Section

FIGURE E6.6   Exercise E6.6.

This problem is somewhat academic, in that the flange width of 30 inches is likely larger than would be used in practice by most engineers, but serves to illustrate the calculations. A more probable solution might be to use lateral bracing in two or three bays adjacent to each support (abutments and piers) to reduce the effective span for torsional brace (cross-frame) design.

E6.7

A welded steel plate girder made from steel with $F_y = 50$ ksi is used for an I-girder bridge. Girder spacing, $S, = 12$ ft, deck thickness, $t_s = 8$ inches, and concrete strength, $f'_c = 4$ ksi. The top flange is 16 inches × 0.8125 inches and the bottom flange is 16 inches × 1.125 inches. The web is 96 inches × 0.6875 inches. The distance from the top of the web to the bottom of the deck is 3 inches. For positive flexure (compression in the top of the deck), determine the plastic moment, M_p, and the yield moment, M_y. Properties (section moduli at the top of the girder and the bottom of the girder) at various stages are provided in Table E6.7. Unfactored moments are:

- $M_{DC1} = 3,260$ ft-kips (carried by the girder alone)
- $M_{DC2} = 661$ ft-kips (carried on the long-term composite section)
- $M_{DW} = 882$ ft-kips (carried on the long-term composite section)

E6.8

A welded steel plate girder made from steel with $F_y = 50$ ksi includes 12 in² of reinforcing in the deck over the interior supports in the negative moment region. The top flange is 20 inches × 2.75 inches and the bottom flange is

TABLE E6.7

Exercise E6.7

Property	Girder	Composite (n)	Composite ($3n$)
I, in^4	122,983	315,538	229,493
S_{top}, in^3	2,396	18,599	7,106
S_{bott}, in^3	2,639	3,897	3,496

22 inches $\times$ 2.75 inches. The web is 96 inches $\times$ 0.6875 inches. Deck reinforcing is located 4.75 inches above the top of the girder and $f_y = 60$ ksi
for the bars. Compute the plastic and yield moments in negative flexure.
Unfactored moments are as follows:

- $M_{DC1} = 9,176$ ft-kips (carried by the girder alone)
- $M_{DC2} = 1,672$ ft-kips (carried on the long-term composite section)
- $M_{DW} = 2,229$ ft-kips (carried on the long-term composite section)

For the girder alone, $I = 331,931$ in^4 and the center of gravity is located
49.25 inches from the bottom of the girder.

E6.9

Suppose the Project Bridge is located at a site characterized by surface
category D conditions for wind load calculations, with $V = 115$ mph for the
Strength III limit state from maps, and height, $Z = 35$ feet. Check the flexural resistance at the Strength I, Strength III, and Strength V limit states for
the maximum negative moment section over the pier. Cross section details
and moments from a Simon model are given below. All steel is Grade 50W
weathering steel. The LTB modification factor, C_b, may be taken to be equal
to 1.36 for all calculations.

- Web plate, 36 inch $\times$ 0.65 inches thick
- Flange plates, 14 inch $\times$ 1.50 inches thick
- Cross-frame spacing adjacent to the pier $= 15$ feet (Figure 1.2)
- Reinforcing bars in deck, $A_s = 10$ in^2, 7.5 inches above the top of the web
- $M_{DC1} = -1,273$ ft-kips, non-composite
- $M_{DC2} = -202$ ft-kips, composite
- $M_{DW} = -283$ ft-kips, composite
- $M_{LL+IM} = -1,366$ ft-kips, composite

7 Precast Prestressed Concrete Girders

Precast, prestressed concrete girders are efficient, economical options for bridges with spans of less than about 150 feet in many locations. Some states, Florida and Washington for example, do have deep precast concrete girders capable of spanning much greater distances. Spliced girder designs have also been used to extend span capabilities for prestressed girders. However, when spans are longer than about 150 feet, steel girders become the preferred choice in areas where the standard AASHTO I-beam and Bulb-T beam cross sections are the only readily available concrete I-girder sections.

The basic idea underlying prestressed girder design is to fully or partially offset the effects of gravity loading with opposing moments and axial loads. At midspan, gravity loads impart tension into the bottom flange of an I-girder. Prestressing applied below the centroid of the girder produces compression in the bottom flange through both a compressive axial force and a negative moment due to the eccentricity of the axial force.

At girder ends made continuous and composite with the concrete deck over interior supports, negative moment reinforcement in the deck, coupled with girder strand extensions into a cast-in-place concrete diaphragm, are relied upon to provide the continuity.

Two of the most commonly used strands are 0.50-inch diameter (area = 0.153 in^2) and 0.60-inch diameter (area = 0.217 in^2) low-relaxation seven-wire strands with tensile strength, f_{pu} = 270 ksi. For these strands, E = 28,500 ksi and f_{py} = 0.90f_{pu}.

Chapter 1 of this text includes a presentation of standard, prestressed girder shapes and properties.

7.1 STRESS ANALYSIS

Figure 7.1 depicts the general stress states from various loading sources on a non-composite precast, prestressed concrete (PPC) girder (prior to it becoming composite with the concrete deck).

AASHTO requirements for PPC girders include Service limit state stress limits in addition to Strength limit state flexural resistance.

Some owners also require that the net deflection due to all load sources be upward, not downward. The requirement is sometimes specified to be satisfied using deflection multipliers equal to 1.0 on all sources, rather than the default values in many software packages.

Stress limits are specified in the AASHTO LRFD BDS both at strand detensioning and for the in-place girder, subject to full design loads. Stress limits at

DOI: 10.1201/9781003265467-7

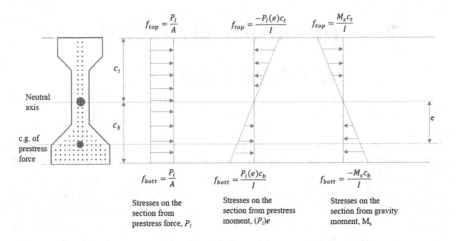

FIGURE 7.1 Precast Girder Stresses.

de-tensioning are based on non-composite girder properties along with the specified minimum girder release concrete strength, f'_{ci}. Final condition stress limits are based on composite section properties and the final required minimum concrete strength of the girder, f'_c.

For temporary stresses prior to losses, concrete compression is limited to $0.65 f'_{ci}$.

For temporary stresses prior to losses, for normal-weight concrete in areas with bonded reinforcement (bars or strands) sufficient to resist concrete tensile force, concrete tension is limited to $0.24\sqrt{f'_{ci}}$.

Final stresses for normal-weight concrete, after all losses have occurred, are limited to the values shown in Table 7.1. The reduction factor, ϕ_w, is applicable to box girders and is equal to 1.0 whenever the flange and web slenderness values are both less than 15. Refer to Section 5.9.2.3.2 of the AASHTO LRFD BDS for other conditions of web and flange slenderness.

For stress calculations at release, one recommended method suggests the use of transformed properties along with the initial pull, with elastic shortening loss effects ignored (Swarz et al., 2012).

7.2 FLEXURAL RESISTANCE

For positive flexure (compression in the bridge deck) at the Strength limit state, Equation 7.1 provides the theoretical nominal resistance for the simplified case in which any mild reinforcement in the tension region of the beam and any compression reinforcement in the slab are both ignored. If the stress block depth, given by Equation 7.3, is less than the deck thickness, then $(b-b_w) = 0$ and the entire second term on the right-hand side of Equation 7.1 is zero. A_{ps} is the total area of strands, f_{ps} is the strand stress at nominal resistance, b is the effective width of the deck, and d_p is the distance from the top of the deck to the centroid of the strands. The deck concrete strength, rather than the girder concrete strength, is to be used in the equations,. The

TABLE 7.1
Final Stress Limits in Prestressed Concrete Bridge I-Girders

Limit State and Loading Conditions	Limiting Stress
Service I limit state compression Effective prestress plus permanent load	$0.45f'_c$
Service I limit state compression Effective prestress, permanent loads, transient loads	$0.60\phi_w f'_c$
Service III limit state tension w/ bonded reinforcement Moderate corrosion conditions	$0.19\sqrt{f'_c} \leq 0.600$ ksi
Service III limit state tension w/ bonded reinforcement Severe corrosion conditions	$0.0948\sqrt{f'_c} \leq 0.300$ ksi

coefficient, k, in the equations is 0.28 for Low-Lax strands. The resistance factor, ϕ, for tension-controlled, prestressed concrete beams is 1.0.

$$M_n = A_{ps}f_{ps}\left(d_p - \frac{a}{2}\right) + \alpha_1 f'_c\left(b - b_w\right)h_f\left(\frac{a - h_f}{2}\right) \tag{7.1}$$

$$c = \frac{A_{ps}f_{pu}}{\alpha_1 f'_c \beta_1 b + kA_{ps}f_{pu}/d_p} \tag{7.2}$$

$$a = \beta_1 c \tag{7.3}$$

$$\alpha_1 = \begin{cases} 0.85, & f'_c \leq 10 \text{ ksi} \\ 0.75, & f'_c \geq 15 \text{ ksi} \\ \text{interpolate for intermediate values} \end{cases} \tag{7.4}$$

$$\beta_1 = \begin{cases} 0.85, & f'_c \leq 4 \text{ ksi} \\ 0.65, & f'_c \geq 8 \text{ ksi} \\ \text{interpolate for intermediate values} \end{cases} \tag{7.5}$$

$$f_{ps} = f_{pu}\left(1 - k \times \frac{c}{d_p}\right) \tag{7.6}$$

$$k = 2\left(1.04 - \frac{f_{py}}{f_{pu}}\right) \tag{7.7}$$

$$\frac{f_{py}}{f_{pu}} = \begin{cases} 0.90, & \text{strands} \\ 0.85, & \text{plain bars} \\ 0.80, & \text{deformed bars} \end{cases} \tag{7.8}$$

For negative flexure over interior piers, a strand stress equal to zero at the girder ends may be used, and negative moment deck reinforcement for the composite beam is designed, using reinforced concrete principles.

7.3 SHEAR RESISTANCE

The summary of shear provisions for prestressed girders presented here is applicable to so-called "B-regions" without significant torsion. These are regions where the assumption that plane sections remain plane is judged to be valid. For deep beams and other disturbed regions ("D-regions") the reader is referred to Section 5.8 of the AASHTO LRFD BDS. Typical, prestressed concrete bridge girders are usually designed as B-regions, provided all requirements in Section 5.7 of the AASHTO LRFD BDS are satisfied.

The shear depth, d_v, is taken to be equal to the distance between the cross-section resultant tensile and compressive forces but need not to be taken to be less than either $0.72h$ or $0.90d_e$; d_e is the distance from the extreme compression fiber to the resultant cross-section tensile force and h is the total member depth. The shear width, b_v, is the minimum web width.

Nominal shear resistance, V_n, consists of contributions from the concrete, from the steel shear reinforcement, and from the vertical component of prestress force at a given section. Equation 7.9 provides the nominal resistance. The units on f'_c must be ksi in the following equations. For normal-weight concrete with stirrups inclined at 90 degrees to the longitudinal axis of the beam, Equations 7.10 and 7.11 provide the concrete and steel contributions, respectively. V_p is the vertical component of prestress force at the section under consideration.

$$V_n = V_c + V_s + V_p \leq 0.25 f'_c b_v d_v \qquad (7.9)$$

$$V_c = 0.0316 \beta \sqrt{f'_c} b_v d_v \qquad (7.10)$$

$$V_s = \frac{A_v f_y d_v}{s} \cdot \cot\theta \qquad (7.11)$$

The minimum area of shear reinforcement is required whenever $V_u > 0.5\phi(V_c + V_p)$ and is given by Equation 7.12 for normal-weight concrete. The resistance factor for shear is 0.85.

$$A_v \geq 0.0316\sqrt{f'_c} \cdot \frac{b_v s}{f_y} \qquad (7.12)$$

Shear reinforcement spacing limits depend on the shear stress on the concrete, v_u, which is calculated using Equation 7.13. The spacing limits are given in Equation 7.14.

$$v_u = \frac{|V_u - \phi V_p|}{\phi b_v d_v} \qquad (7.13)$$

$$s_{max} = \begin{cases} 0.4d_v \le 12 \text{ inches,} & \text{if } v_u \ge 0.125 f_c' \\ 0.8d_v \le 24 \text{ inches,} & \text{if } v_u < 0.125 f_c' \end{cases} \quad (7.14)$$

Calculation of the coefficients β and θ is required to accurately assess shear resistance. These factors depend on the net tensile strain in the centroid of the tension reinforcement, ε_s. The discussion here is limited to sections with at least the minimum amount of transverse shear reinforcement. For such cases, the strain and the coefficients are given by Equations 7.15 through 7.17. The strain may also be determined by a cross-sectional analysis. For other cases, the reader is referred to Section 5.7.3.4.2 of the AASHTO LRFD BDS.

$$\varepsilon_s = \frac{\left(\dfrac{|M_u|}{d_v} + 0.5N_u + |V_u - V_p| - A_{ps}f_{po} \right)}{E_s A_s + E_p A_{ps}} \quad (7.15)$$

$$\beta = \frac{4.8}{1 + 750\varepsilon_s} \quad (7.16)$$

$$\theta = 29 + 3500\varepsilon_s \quad (7.17)$$

For typical values of prestressing parameters, f_{po} may be taken to be equal to zero at the point of bonding between strands and surrounding concrete up to $0.7f_{pu}$ at the transfer length. Refer to Section 5.7.3.4.2 of the AASHTO LRFD BDS for additional discussion on f_{po}.

For calculated strains, ε_s, of less than zero, ε_s may be taken to be equal to zero. The upper limit on ε_s to be used for calculation of shear resistance coefficients is $\varepsilon_s \le 0.006$. This effectively places a lower limit on β equal to 0.873 and an upper limit on θ equal to 50 degrees. Other limitations on the parameters are summarized below.

- N_u, the factored axial force on the cross section, is positive if tension is applied, negative if compression is applied.
- M_u, the factored moment at the section, is not to be less than $(V_u - V_p) d_v$.
- For sections closer than d_v to the face of support, ε_s, at a distance d_v from the face of support, may be used to determine β and θ, unless a concentrated load is located within d_v from the support, in which case ε_s should be determined at the face of support.

For additional shear-related longitudinal reinforcement requirements, refer to Section 5.7.3.5 of the AASHTO LRFD BDS.

Section 5.7.4 of the AASHTO LRFD BDS covers interface shear reinforcement requirements. Regarding prestressed bridge girders, the provisions are applicable to horizontal shear transfer between the precast, prestressed girder and the cast-in-place concrete deck. For such members, interface shear transfer is typically accomplished by extending a portion of the girder shear stirrups into the deck with terminating hooks.

The minimum required area of interface shear reinforcement is given by Equation 7.18. The value for f_y in Equations 7.18 and 7.19 is not to exceed 60 ksi, regardless of the reinforcement grade used for shear steel. The design interface shear resistance is given by Equation 7.19. P_c is the permanent net compressive force normal to the shear plane.

$$A_{vf} = \frac{0.05A_{cv}}{f_y} \tag{7.18}$$

$$\phi V_{ni} = \phi \left(cA_{cv} + \mu A_{vf} f_y + \mu P_c \right) \le \phi K_1 f_c' A_{cv} \le \phi K_2 A_{cv} \tag{7.19}$$

Table 7.2 summarizes μ, c, K_1, and K_2 for various conditions using normal-weight concretes. For conditions with lightweight concretes, refer to Section 5.7.4.4 of the AASHTO LRFD BDS.

The maximum spacing of interface shear transfer reinforcement is limited to 48 inches or the girder depth, whichever is smaller.

7.4 CONTINUITY DETAILS

Prestressed girders are typically constructed to span between successive supports (abutments or piers) as simply supported beams for self-weight, wet deck concrete weight, intermediate diaphragms, and construction loads prior to deck concrete curing. Prestressed concrete girders running continuously over interior supports require longitudinal deck reinforcement, cast-in-place support diaphragms, and the extension of girder strands into the diaphragm to enable the composite girder to resist negative moments. Refer to the LRFD BDS, Section 5.12.3.3, for additional details on continuity not fully presented here.

TABLE 7.2
Interface Shear Parameters

Conditions	μ	c	K_1	K_2
Cast-in-place concrete slab on clean concrete girder surfaces, free of laitance, with surface intentionally roughened to an amplitude of 0.25 inches	1.0	0.28 ksi	0.30	1.8 ksi
Normal-weight concrete placed monolithically	1.4	0.40 ksi	0.25	1.5 ksi
Normal-weight concrete placed against a clean concrete surface, free of laitance, with surface intentionally roughened to an amplitude of 0.25 inches	1.0	0.24 ksi	0.25	1.5 ksi
Concrete placed against a clean concrete surface, free of laitance, but not intentionally roughened	0.6	0.075 ksi	0.20	0.80 ksi
Concrete anchored to as-rolled structural steel by headed studs or by reinforcing bars, where all steel in contact with concrete is clean and free of paint	0.7	0.025 ksi	0.20	0.80 ksi

Restraint moments from time-dependent deformations develop in precast girders made continuous. These restraint moments can be ignored in design as long as the girder age, at the time continuity is achieved, is 90 days or more. For other cases, refer to the LRFD BDS, Section 5.12.3.3.2 and Commentary, for restraint moment calculations to be accounted for in the design. Restraint moments may be positive or negative. Positive restraint moments at girder ends may reduce the effectiveness of the continuity diaphragm by producing excessive cracking at the compression block. Hence, if restraint moments are neglected in the girder design, a plans-note stating that the girder age must be no less than 90 days at the time continuity is established will be required.

Even when girder age is specified on the plans to be no less than 90 days at continuity, the LRFD BDS requires that a positive moment connection with a design resistance, ϕM_n, no less than 1.2 times the cracking moment, M_{cr}, be provided. Such positive moment connections may consist of at least three options.

1. Non-prestressed reinforcement embedded in the girder and developed into the continuity diaphragm
2. Extension of prestressing strands beyond the end of the girder and anchored into the continuity diaphragm.
3. Testing to validate other methods an owner may propose.

Strands de-bonded at girder ends to control tensile stresses in the girder are not to be used as elements contributing to the positive moment connection.

For normal-weight concrete with specified strength of no more than 15 ksi, the cracking moment, $M_{cr} = 0.24(f'_c)^{0.5}(I_g)/y_t$, is to be computed using the composite girder/deck geometric properties, but with a concrete strength corresponding to that of the continuity diaphragm, which is generally cast simultaneously with the deck over the support.

When projecting strands are used for the positive moment connection, the LRFD BDS, in Sections 5.12.3.3.9c and 5.12.3.3.9d, requires that several criteria be satisfied.

- Extended strands must be anchored into the continuity diaphragm with either 90-degree hooks or a strand development length.
- Extended strands must extend no less than 8 inches from the end of the girder prior to bending.
- Extended strands shall form a generally symmetrical pattern about the centerline of the girder.
- Strands from opposing girders must be detailed so as to preclude conflicts in the reinforcement patterns.

For stress in the extended strands, Equations 7.20 (for the Service limit state) and 7.21 (for the Strength limit state) from the LRFD BDS Commentary can be used. Equations 7.20 and 7.21 were developed based on 0.5-inch-diameter strands. Although not in the LRFD BDS, equations for 0.6-inch-diameter strands used in the original report (Miller, et al., 2004) for continuity calculations are presented here as

Equations 7.22 and 7.23, based on development lengths proportional to strand diameters. The total length of the extended strand, l_{dsh}, includes both the straight and bent portions of the strand. The resulting calculated stresses are in ksi.

$$f_{psl} = \frac{(l_{dsh} - 8)}{0.228}, \; 0.50\text{-inch strand} \tag{7.20}$$

$$f_{pul} = \frac{(l_{dsh} - 8)}{0.163}, \; 0.50\text{-inch strand} \tag{7.21}$$

$$f_{psl} = \frac{\left(l_{dsh} \left(\dfrac{0.50}{0.60} \right) - 8 \right)}{0.228}, \; 0.60\text{-inch strand} \tag{7.22}$$

$$f_{pul} = \frac{\left(l_{dsh} \left(\dfrac{0.50}{0.60} \right) - 8 \right)}{0.163}, \; 0.60\text{-inch strand} \tag{7.23}$$

With girder age no less than 90 days at the time continuity is established, the positive connection design moments may be taken to be equal to M_{cr} at the Service limit state and $1.2M_{cr}$ at the Strength limit state. Although not explicitly stated in the AASHTO LRFD BDS, these design moments were used in the research upon which the LRFD BDS provisions are based.

The original research (Miller et al., 2004) upon which the LRFD BDS provisions are based recommends a total strand embedment, l_{dsh}, such as to produce a Service limit state stress, f_{psl}, no greater than 150 psi. This results in a 42-inch embedment for 0.5-inch-diameter strands and a 51-inch embedment for 0.60-inch-diameter strands. Of course, when analysis indicates that a shorter embedment satisfies all requirements, a shorter embedment may be used. Little, if any, benefit is gained by using larger than necessary embedment, it seems.

7.5 MILD TENSILE REINFORCEMENT IN GIRDERS

Prestressed concrete I-girders, whether continuous or not for live loads, typically require non-prestressed reinforcement in the top flange of the girder to resist tensile stresses at strand release.

The required area of steel in the girder top flange may be calculated using the method proposed in the LRFD BDS Commentary to Section 5.9.2.3.1b. The stresses required for the calculations are those in the top, f_{citop}, and bottom, f_{cibot}, of the girder at strand release prior to losses in strand stress. With reference to Figure 7.1, these stresses at the girder ends can be approximated to by Equations 7.24 and 7.25, with the required area of steel, A_s, given by equation 7.26. The parameter, b_{top}, is the top flange width and x is the distance from the top of the girder to the location of zero stress in the girder. The total height of the non-composite girder is h. Cross-sectional properties of the girder include the area, A_g, the moment of inertia, I_g, and the distances from the neutral axis to the top, c_t, and bottom, c_b, of the girder. P_i is the total, initial, prestress force and e is the distance from the centroid of the strands to the girder neutral axis.

$$f_{citop} = \frac{P_i}{A_g} - \frac{(P_i \times e)c_t}{I_g} \tag{7.24}$$

$$f_{cibot} = \frac{P_i}{A_g} + \frac{(P_i \times e)c_b}{I_g} \tag{7.25}$$

$$A_s = \frac{f_{citop}b_{top}x}{2f_s} \tag{7.26}$$

$$x = \left(\frac{f_{citop}}{f_{cibot} - f_{citop}} \right)h \tag{7.27}$$

$$f_s = 0.50f_y \leq 30 \text{ ksi} \tag{7.28}$$

The sign convention used here is negative for tensile stress. Should the calculated stress in the top of the girder, f_{citop}, be positive (compression), then the theoretical area of mild reinforcement required is zero. The sign on the stresses calculated should be retained to properly evaluate the equations.

7.6   NEGATIVE MOMENT REINFORCEMENT FOR GIRDERS MADE CONTINUOUS

For precast, prestressed girders made continuous for composite dead load and live load, deck reinforcement is required to carry tension in the regions for which negative moment exists in the continuous beam. Such reinforcement is typically epoxy coated and may consist of longitudinal bars in the deck in two mats. Careful attention to bar size is necessary, given the congestion possible in relatively thin bridge decks with two mats of steel running in each direction.

Some states permit the use of partial depth, precast panels for bridge decks, at the contractor's option. In such cases, the contractor is sometimes permitted to omit the bottom mat of longitudinal reinforcing in the deck. Whenever the bottom mat is counted on as part of the negative moment reinforcing, however, such omission must not be permitted. The engineer must either use only the top mat of longitudinal bars to carry the negative moment tension, or disallow the use of precast panels.

Negative moment reinforcement in the deck is subject to Strength, Service, and Fatigue limit state requirements in the AASHTO LRFD BDS.

Strength limit state flexural resistance may be computed using Equation 7.29, based on the simplifying assumptions that (a) the effects of any mild, compression reinforcement in the bottom of the beam may be ignored, and (b) the tension steel has yielded (this must always be verified, or adjustment made to the calculated resistance).

The flange thickness, h_f, is that for the bottom of the prestressed beam. The flange width, b, is for the bottom of the prestressed beam, and the web width, b_w, is for the prestressed beam. The depth from the extreme compression fiber (at the bottom of the beam) to the centroid of the negative moment reinforcing in the deck is d_s. If the calculated stress block depth, a, is less than h_f then the entire second term on the right-hand side of Equation 7.29 is zero, as is the second term in the right-hand side numerator of Equation 7.30, since $(b-b_w)$ is zero in such a case.

Given that the maximum negative moment occurs at the centerline of the interior supports (piers or bents), one strategy is to incorporate in the equations the compressive strength of the cast-in-place diaphragm (which is typically the same concrete as used for the deck and is cast simultaneously with the deck) rather than that for the beam. This is recommended practice, even though the bottom of the beam is under compression, since the compressive stress block also exists in the cast-in-place diaphragm (at least at the end of the girder).

$$\phi M_n = \phi \left[A_s f_y \left(d_s - \frac{a}{2} \right) + \alpha_1 f_c' (b - b_w) h_f \left(\frac{a - h_f}{2} \right) \right] \tag{7.29}$$

$$c = \frac{A_s f_y - \alpha_1 f_c' (b - b_w) h_f}{\alpha_1 f_c' \beta_1 b_w} \tag{7.30}$$

$$a = \beta_1 c \tag{7.31}$$

$$\alpha_1 = \begin{cases} 0.85, & f_c' \le 10 \text{ ksi} \\ 0.75, & f_c' \ge 15 \text{ ksi} \\ \text{interpolate for intermediate values} \end{cases} \tag{7.32}$$

$$\beta_1 = \begin{cases} 0.85, & f_c' \le 4 \text{ ksi} \\ 0.65, & f_c' \ge 8 \text{ ksi} \\ \text{interpolate for intermediate values} \end{cases} \tag{7.33}$$

Negative moment, deck reinforcement design for precast beams made continuous often results in a tension-controlled section, with a corresponding resistance factor (ϕ) equal to 0.90. This must be verified by checking the strain in the tension reinforcement, and if the section is not tension-controlled, then the resistance factor must be reduced. The resistance factor for compression-controlled sections is 0.75. For sections neither tension controlled nor compression controlled, linear interpolation is used to determine the appropriate ϕ factor.

Tension-controlled reinforced concrete sections are defined as sections with a net tensile strain in the extreme layer of tensile reinforcement, ε_t, greater than or equal to the tension-controlled strain limit, ε_{tl}, when the concrete strain reaches a value of 0.003.

Compression-controlled reinforced concrete sections are defined as sections with a net tensile strain in the extreme layer of tensile reinforcement, ε_t, less than or equal to the compression-controlled strain limit, ε_{cl}, when the concrete strain reaches a value of 0.003.

The tension-controlled strain limit, ε_{tl}, is determined as follows:

- $\varepsilon_{tl} = 0.005$ for reinforcement with $f_y \leq 75$ ksi
- $\varepsilon_{tl} = 0.008$ for reinforcement with $f_y = 100$ ksi
- ε_{tl} is determined by linear interpolation for reinforcement with $75 < f_y < 100$ ksi

The compression-controlled strain limit, ε_{cl}, is determined as follows:

- $\varepsilon_{cl} = f_y / E_s$, but not > 0.002, for reinforcement with $f_y \leq 60$ ksi
- $\varepsilon_{cl} = 0.004$ for reinforcement with $f_y = 100$ ksi
- ε_{cl} is determined by linear interpolation for reinforcement with $60 < f_y < 100$ ksi

For the Service limit state, negative moment deck reinforcing is subject to the crack control requirements of the AASHTO LRFD BDS, Section 5.6.7. The lateral spacing (s) of longitudinal bars must not exceed the value given by Equation 7.34. The exposure factor, γ_e, is 1.0 for Class 1 exposure conditions and 0.75 for Class 2 exposure conditions. Decks are typically assigned Class 2 exposure due to their susceptibility to corrosive conditions. Some owners choose to adopt an intermediate value for γ_e.

The parameter, β_s, is the ratio of the distance from the cracked neutral axis to the extreme tension face to the distance from the cracked neutral axis to the centroid of reinforcement closest to that face. The actual definition involves strains, but, under the assumption that strains are proportional to distance from the neutral axis, the definition provided is equivalent. An estimate of the value may be calculated from Equation 7.35, or a more precise value may be obtained from strain compatibility analysis.

The stress, f_{ss}, in the tension steel at the Service limit state, is to be computed based on cracked section properties, and is not to exceed $0.6f_y$. The overall height of the member is h, and the distance from the extreme tension face to the centroid of the reinforcement closest to that face is d_c.

$$s \leq \frac{700\gamma_e}{\beta_s f_{ss}} - 2d_c \qquad (7.34)$$

$$\beta_s = 1 + \frac{d_c}{0.7(h - d_c)} \qquad (7.35)$$

Although Section 5.5.3 of the AASHTO LRFD BDS specifically excludes deck slabs from fatigue criterion requirements, this exception presumably applies only to the design of the transverse deck reinforcement, given the Commentary note that the exclusion of decks from fatigue requirements is based on the arching action of the deck. The behavior of a composite beam is very different from that of a bridge deck, and such arching is not typically present in bridge girders.

For sections at which the unfactored dead load stress sum is less than that resulting from the Fatigue I (infinite life) limit state, fatigue requirements are satisfied by limiting the stress range in the deck bars to that given by Equation 7.36, as long as no welds or splices are present at the section under consideration. The stress, f_{min}, is the minimum stress resulting from (a) the Fatigue I limit state combined with (b) the unfactored dead load effects (both DC and DW). Only those DC and DW effects which act on the composite section, after continuity has been established, should be included. The stress, f_{min}, is positive if tension, negative if compression.

$$\gamma\left(\Delta f\right) = 1.75\left(\Delta f\right) \le \left(\Delta F\right)_{TH} = 26 - 22\frac{f_{min}}{f_y} \tag{7.36}$$

If negative moment reinforcement is spliced, then the limiting stress range $(\Delta F)_{TH}$ is reduced significantly to 4.0 ksi.

7.7 TRANSFER AND DEVELOPMENT LENGTH

Transfer length is the distance from the stress-free unbonded end of a strand to the point at which the strand stress reaches a value equal to the effective prestress, f_{pe}, after losses.

The calculation of losses in pretensioned members is not covered in this text. Modern software typically contains multiple options for the calculation of prestress losses. Refer to Section 5.9.3 of the AASHTO LRFD BDS.

Development length is defined as the distance from the stress-free, unbonded end of a strand to the point at which strand stress reaches a value equal to the stress at nominal resistance, f_{ps}.

Transfer length is equal to 60 strand diameters, and development length is equal to the value given by Equation 7.37. The nominal strand diameter is d_b. The coefficient, κ, is equal to 1.6 for pretensioned beams with a depth greater than 24 inches, or 1.0 for other pretensioned members.

$$l_d = \kappa\left(f_{ps} - \frac{2}{3}f_{pe}\right)d_b \tag{7.37}$$

7.8 STRESS CONTROL MEASURES

Controlling stresses in pretensioned bridge girders, both at release and for in-service conditions, may require one or more of the following:

- draping of strands
- debonding of strands
- increase in the compressive strengths specified for the beam

Draping, also referred to as harping, involves raising strands at the end of the beam relative to the location at the midspan of the beam in an effort to mitigate over-stresses. Draping effectively raises the centroid of the prestressing strands near the end of the beam. The onset of the drape is typically about 40 percent of the beam length from each end, and requires special hold-down devices to resist the upward component of strand force. Figure 7.2 shows the elevation of a prestressed girder incorporating draped strands to control stresses near the end of the beam. Figure 7.3 shows the mid-span and girder end cross sections for the same girder.

De-bonding involves wrapping strands, typically near the end of the beam, to prevent bonding to the concrete until such point along the beam as such bonding does not overstress the beam.

The AASHTO LRFD BDS is valid for concrete strengths up to 15 ksi. Regional practices may make it difficult to obtain concrete strengths near 15 ksi, and values closer to 10 ksi are often targeted by engineers. Recall that specified strengths include that at release, f'_{ci}, and that for in-service conditions, f'_c. In-service required strength is typically of the order of 500 to 1,000 psi greater than the required strength specified at release, though no such range is a requirement.

7.9   SOLVED PROBLEMS

Problem 7.1

For the concrete girder option of the Project Bridge, the 89-foot 3-inch-long BT-54 girder shown in Figure P7.1 below has been proposed. The section at midspan is shown in the figure. Prestressing steel includes thirty ½-inch-diameter, 270K low-relaxation strands, arranged as denoted by the solid circles in the cross-section figure. Deck thickness is 8.25 inches and the girder spacing is 9 ft 3 inches. Deck concrete f'_c is 4.0 ksi. Girder release and final strengths are $f'_{ci} = 7.0$ ksi and $f'_c = 9.2$ ksi. The bottom row of

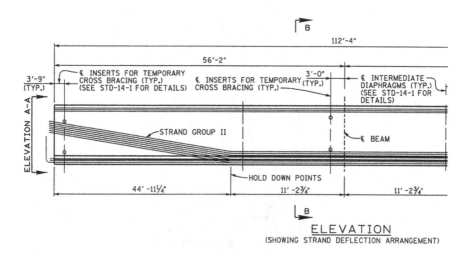

FIGURE 7.2   Partial Girder Elevation Showing Draped Strands.

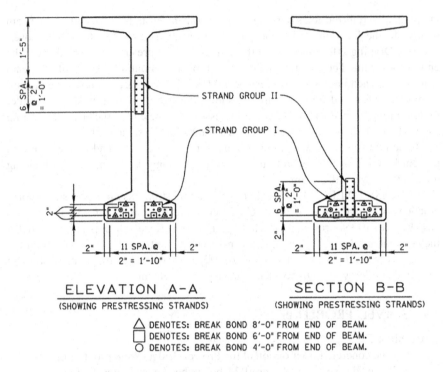

FIGURE 7.3 Cross Sections Showing Draped Strands.

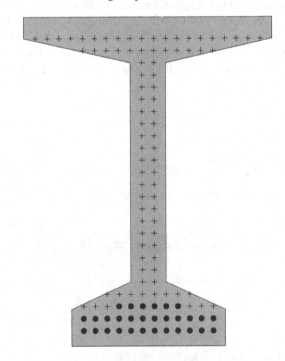

FIGURE P7.1 Problem P7.1.

strands is centered 2.5 inches from the bottom of the girder. Subsequent rows are 2 inches above the previous row. The haunch distance, from the top of girder to the bottom of deck, is 2 inches. The deck reinforcement over the intermediate pier is 22 #7 bars (f_y = 60 ksi). Assume the centroid of the 22 bars is at mid-thickness of the deck.

Determine the initial pull and the stresses in the girder at release. Compare the computed stresses to those permitted at release. Use transformed properties and ignore losses due to elastic shortening for stress calculations.

Problem 7.2

Maximum midspan positive (tension in the bottom of the girder) girder shears and moments for the composite BT-54 in Problem 7.1 are as follows: case 1 (maximum positive moment)

$$M_{DC} = 1,985 \text{ ft-kips} \quad M_{DW} = 143 \text{ ft-kips} \quad M_{LL+IM} = 1,421 \text{ ft-kips}$$

$$V_{DC} = 3 \text{ kips} \quad V_{DW} = 5 \text{ kips} \quad V_{LL+IM} = 33 \text{ kips}$$

case 2 (maximum shear)

$$M_{DC} = 1,985 \text{ ft-kips} \quad M_{DW} = 143 \text{ ft-kips} \quad M_{LL+IM} = 1,103 \text{ ft-kips}$$

$$V_{DC} = 3 \text{ kips} \quad V_{DW} = 5 \text{ kips} \quad V_{LL+IM} = 44 \text{ kips}$$

Determine the required flexural resistance at the Strength limit state, M_u, and the flexural resistance, ϕM_n, in the design provided. The flexural resistance is from the thirty ½-inch-diameter strands in tension and the deck in compression. Use hand calculations supplemented with a Response 2000 model. Total losses are estimated to be 19.3 ksi. The deck reinforcement is to be ignored. Shear stirrups are single #6 at 18 inches.

Problem 7.3

Maximum negative (compression in the bottom of the girder) girder moments and corresponding shears for the composite BT-54 in Problem 7.1 are as follows:

case 1 (maximum negative moment)

$$M_{DC} = -192 \text{ ft-kips} \quad M_{DW} = -269 \text{ ft-kips} \quad M_{LL+IM} = -1,519 \text{ ft-kips}$$

$$V_{DC} = 94 \text{ kips} \quad V_{DW} = 16 \text{ kips} \quad V_{LL+IM} = 84 \text{ kips}$$

case 2 (maximum shear)

$$M_{DC} = -192 \text{ ft-kips} \quad M_{DW} = -269 \text{ ft-kips} \quad M_{LL+IM} = -612 \text{ ft-kips}$$

$$V_{DC} = 94 \text{ kips} \quad V_{DW} = 16 \text{ kips} \quad V_{LL+IM} = 112 \text{ kips}$$

Determine the required flexural resistance at the Strength limit state, M_u, and the flexural resistance, ϕM_n, of the provided design. The flexural resistance provided is carried by the bars in the deck in tension. Stirrups are double #6 at 6 inches on center extending into the deck.

Problem 7.4

Estimate the midspan deflection of the BT-54 girder from Problem 7.1. The effects to be included are prestress and girder self-weight.

Problem 7.5

Determine the continuity requirements for strand extension into the support diaphragm at the intermediate pier of the two-span Project Bridge.

PROBLEM 7.1	TEH	1/3

$$BT\text{-}54: \quad A = 659 \text{ in}^2$$

$$I_x = 268,077 \text{ in}^4$$

$$y_b = 27.63 \text{ in.}$$

$$wt. = 0.686 \text{ KLF}$$

Strand Centroid:

$$\overline{y}_{strands} = \frac{12(2.5 + 4.5) + 6(6.5)}{30}$$

$$\Rightarrow \overline{y}_{strands} = 4.10 \text{ inches from bottom of girder}$$

$$A_{ps} = 30 \times 0.153 = 4.59 \text{ in}^2$$

$$P_i = 0.75 \times 270 \times 4.59$$

$$\Rightarrow \underline{P_i = 929.5 \text{ KIPS}}$$

$$E_c = 1,820\sqrt{7} = 4,815 \text{ KSi}$$

$$n = 28,500 / 4,815 = 5.92$$

$$(n-1)A_{ps} = 4.92 \times 4.59 = 22.6 \text{ in}^2$$

Transformed Properties.

$$A' = 659 + 22.6 = 681.6 \text{ in}^2$$

$$y' = \frac{659(27.63) + 22.6(4.10)}{681.6}$$

$$\Rightarrow y' = 26.85 \text{ inches from bottom of girder}$$

PROBLEM 7.1	TEH	2/3

$$I' = 268,077 + 22.6(4.1 - 26.85)^2$$
$$+ 659(27.63 - 26.85)^2$$

$$\Rightarrow I' = 280,175 \text{ in}^4$$

$$S_{top} = \frac{280,175}{54 - 26.85} = 10,319 \text{ in}^3$$

$$S_{bott} = \frac{280,175}{26.85} = 10,435 \text{ in}^3$$

$$e = 26.85 - 4.1 = 22.75''$$

$$M_{ps} = 929.5 \times 22.75 = 21,146 \text{ in·k}$$
$$(\text{moment from prestress})$$

$$M_{sw} = 0.686(89.25)^2/8$$
$$= 683 \text{ ft·k} = 8,197 \text{ in·k}$$
$$(\text{moment from self·weight})$$

$$f_{top} = \frac{929.5}{681.6} + \frac{(8,197 - 21,146)}{10,319}$$

$$\Rightarrow f_{top} = 1.364 - 1.255$$
$$= 0.109 \text{ ksi (compr.)}$$

$$f_{bott} = \frac{929.5}{681.6} - \frac{(8,197 - 21,146)}{10,435}$$

$$\Rightarrow f_{bott} = 1.364 + 1.241$$
$$= 2.605 \text{ ksi (compr)}$$

PROBLEM 7.1	TEH	3/3

Compression limit at release.

$$0.65 \, f'_{ci} = 0.65 \times 7 = 4.55 \text{ ksi}$$

$$> 2.605 \text{ ksi}$$

$$\Rightarrow \underline{OK}$$

Check tension in top near end:

$$f_{top} = \frac{929.5}{681.6} - \frac{21,146}{10,319}$$

$$= -0.685 \text{ ksi (tension)}$$

Tension limit at release.

$$0.24\sqrt{7} = 0.635 \text{ ksi}$$

$$0.685 > 0.635 \Rightarrow \underline{\underline{No \ Good}}$$

change f'_{ci}:

$$.24\sqrt{f'_{ci}} = 0.685$$

$$\Rightarrow f'_{ci} = 8.15 \text{ ksi}$$

PROBLEM 7.2	TEH	1/4

Hand calculation ignoring the M–V interaction:

From Problem 7-1,

$$\overline{y}_{strands} = 4.10'' \text{ from the bottom of the girder}$$

$$d_p = 54 + 2 + 8.25 - 4.10 = 60.15''$$

$$\beta_1 = 0.85 \quad (f'_c = 4 \text{ ksi for deck})$$

$$\alpha_1 = 0.85$$

$$C = \frac{4.59 \times 270}{.85(4)(.85)(9.25\times12) + 0.28(4.59)(270)/60.15}$$

$$\Rightarrow C = 3.79 \text{ inches} < 8.25''$$

$$\text{(assumptions ok)}$$

$$a = .85 \times 3.79 = 3.23''$$

$$M_n = 4.59 \, f_{ps} \left(60.15 - \frac{3.23}{2}\right)/12$$

$$= 22.39 \, f_{ps}$$

$$f_{ps} = 270\left(1 - .28 \times 3.79/60.15\right)$$

$$= 265.2 \text{ ksi}$$

$$\Rightarrow M_n = 5,939 \text{ ft-kips}$$

$$M_u = 1.25(1,985) + 1.50(143) + 1.75(1,421)$$

$$\Rightarrow M_u = 5,183 \text{ ft-k} < 5,939 \text{ ft-k}$$

$$\Rightarrow \underline{ok}$$

PROBLEM 7.2	TEH	2/4

Initial Effective Prestrain :

$$\varepsilon_i = \frac{.75 \times 270 - 19.3}{28,500} = 0.00643$$

$M_u = 1.25(1,985) + 1.50(143) + 1.75(1,421)$

$\Rightarrow M_u = 5,183 \text{ ft·k}$, Case 1

$V_u = 1.25(3) + 1.5(5) + 1.75(33)$

$\Rightarrow V_u = 69 \text{ kips}$, Case 1

$M_u / V_u = 5,183 / 69 = 75.1 \text{ ft}$, Case 1

$M_u = 1.25(1,985) + 1.50(143) + 1.75(1,103)$

$\Rightarrow M_u = 4,626 \text{ ft·k}$, Case 2

$V_u = 1.25(3) + 1.50(5) + 1.75(44)$

$\Rightarrow V_u = 88.3 \text{ kips}$, Case 2

$M_u / V_u = 4,626 / 88.3 = 52.4 \text{ ft}$, Case 2

Response 2000 Model "CEE4380-P07.01"

$M/V = 75.1' \Rightarrow M_n = 5,757 \text{ ft·kips}$
tension-controlled, $\phi = 1.0$
$\phi M_n = 5,757 \text{ ft·k} > M_u = 5,183$
<u>ok</u>

$M/V = 52.4 \Rightarrow M_n = 5,701 \text{ ft·k}$
$\phi M_n = 5,701 > 4,626$ <u>ok</u>

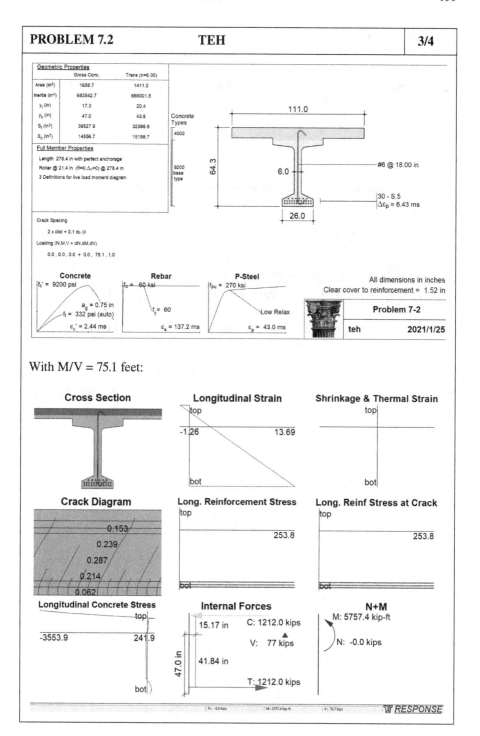

PROBLEM 7.2	TEH		3/4

Geometric Properties

	Gross Conc.	Trans (n=6.00)
Area (in²)	1658.7	1411.2
Inertia (in⁴)	683542.7	666001.5
y_t (in)	17.3	20.4
y_b (in)	47.0	43.8
S_t (in³)	39527.9	32598.6
S_b (in³)	14556.7	15198.7

Full Member Properties

Length: 278.4 in with perfect anchorage
Roller @ 21.4 in (θ=0,Δ_y≠0) @ 278.4 in
3 Definitions for live load moment diagram

Concrete Types
4000
9200 base type

Crack Spacing

2 x dist + 0.1 d_b /p

Loading (N,M,V + dN,dM,dV)

0.0 , 0.0 , 0.0 + 0.0 , 75.1 , 1.0

111.0
64.3
6.0 — #6 @ 18.00 in
30 - S.5
$\Delta\varepsilon_p$ = 6.43 ms
26.0

Concrete
f_c' = 9200 psi
a_g = 0.75 in
f_t = 332 psi (auto)
ε_c' = 2.44 ms

Rebar
f_u = 60 ksi
f_y = 60
ε_s = 137.2 ms

P-Steel
f_{pu} = 270 ksi
Low Relax
ε_p = 43.0 ms

All dimensions in inches
Clear cover to reinforcement = 1.52 in

Problem 7-2
teh   2021/1/25

With M/V = 75.1 feet:

Cross Section

Longitudinal Strain
top
-1.26   13.69
bot

Shrinkage & Thermal Strain
top
bot

Crack Diagram
0.153
0.239
0.287
0.214
0.062

Long. Reinforcement Stress
top
253.8
bot

Long. Reinf Stress at Crack
top
253.8
bot

Longitudinal Concrete Stress
top
-3553.9   241.9
bot

Internal Forces
15.17 in   C: 1212.0 kips
V: 77 kips
47.0 in
41.84 in
T: 1212.0 kips

N+M
M: 5757.4 kip-ft
N: -0.0 kips

N : -0.0 kips   M : 5757.4 kip-ft   V : 76.7 kips   **RESPONSE**

| PROBLEM 7.2 | TEH | 4/4 |

With M/V = 52.4 feet:

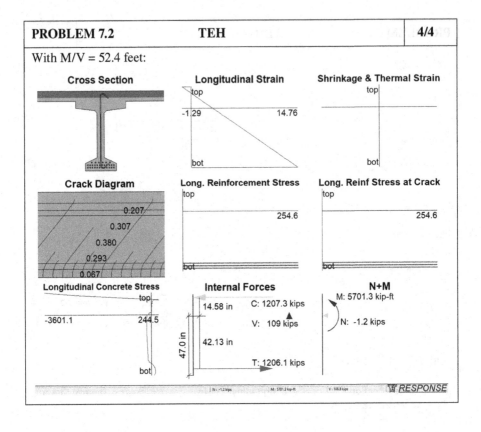

Cross Section

Longitudinal Strain
top
-1.29   14.76
bot

Shrinkage & Thermal Strain
top
bot

Crack Diagram
0.207
0.307
0.380
0.293
0.067

Long. Reinforcement Stress
top
254.6
bot

Long. Reinf Stress at Crack
top
254.6
bot

Longitudinal Concrete Stress
top
-3601.1   244.5
bot

Internal Forces
14.58 in   C: 1207.3 kips
47.0 in   V: 109 kips
42.13 in
T: 1206.1 kips

N+M
M: 5701.3 kip-ft
N: -1.2 kips

N : -1.2 kips   M : 5701.3 kip-ft   V : 105.5 kips   *RESPONSE*

PROBLEM 7.3	TEH	1/3

Case 1:

$$M_u = 1.25(192) + 1.5(269) + 1.75(1519)$$
$$\Rightarrow M_u = 3{,}302 \; ft \cdot k$$
$$V_u = 1.25(94) + 1.5(16) + 1.75(84)$$
$$\Rightarrow V_u = 289 \; kips$$
$$M_u/V_u = 3{,}302/289 = 11.43 \, ft$$

Case 2:

$$M_u = 1.25(192) + 1.5(269) + 1.75(612)$$
$$\Rightarrow M_u = 1{,}714 \; ft \cdot kips$$
$$V_u = 1.25(94) + 1.5(16) + 1.75(112)$$
$$\Rightarrow V_u = 338 \; kips$$
$$M_u/V_u = 1{,}714/338 = 5.07 \, ft$$

Response 2000 Model:
   "CEE4380 - P07.03"

$$M/V = -11.43' \Rightarrow M_n = 3{,}196 \; ft \cdot k < 3{,}302$$
$$M/V = -5.07' \Rightarrow M_n = 2{,}194 \; ft \cdot k > 1{,}714$$

   No Good for Case 1

(change stirrups to double #6 @ 3")
   $\Rightarrow M_n = 3{,}326 \; ft \cdot k > 3{,}302, \, ok$

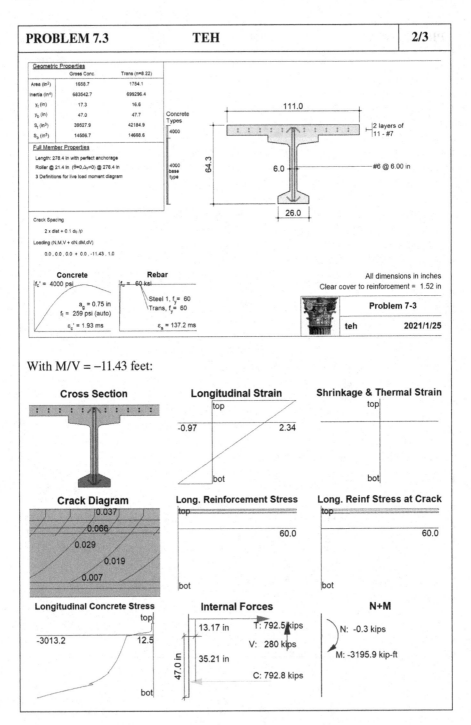

Geometric Properties

	Gross Conc.	Trans (n=8.22)
Area (in²)	1658.7	1754.1
Inertia (in⁴)	683542.7	699296.4
y_t (in)	17.3	16.6
y_b (in)	47.0	47.7
S_t (in³)	39527.9	42184.9
S_b (in³)	14556.7	14668.6

Full Member Properties

Length: 278.4 in with perfect anchorage

Roller @ 21.4 in (θ=0,Δy=0) @ 276.4 in

3 Definitions for live load moment diagram

Concrete Types
4000

4000 base type

111.0

64.3

6.0

26.0

2 layers of
11 - #7

#6 @ 6.00 in

Crack Spacing

2 x dist + 0.1 dc /ρ

Loading (N,M,V + dN,dM,dV)

0.0 , 0.0 , 0.0 + 0.0 , -11.43 , 1.0

Concrete
f_c' = 4000 psi

a_g = 0.75 in
f_t = 259 psi (auto)

ε_c' = 1.93 ms

Rebar
f_u = 60 ksi

Steel 1, f_y= 60
Trans, f_y= 60

ε_s = 137.2 ms

All dimensions in inches

Clear cover to reinforcement = 1.52 in

Problem 7-3	
teh	2021/1/25

With M/V = −11.43 feet:

Cross Section

Longitudinal Strain

top

−0.97   2.34

bot

Shrinkage & Thermal Strain

top

bot

Crack Diagram

0.037

0.066

0.029

0.019

0.007

Long. Reinforcement Stress

top

60.0

bot

Long. Reinf Stress at Crack

top

60.0

bot

Longitudinal Concrete Stress

top

−3013.2   12.5

bot

Internal Forces

13.17 in

35.21 in

47.0 in

T: 792.5 kips

V: 280 kips

C: 792.8 kips

N+M

N: -0.3 kips

M: -3195.9 kip-ft

PROBLEM 7.3	TEH	3/3

With M/V = −5.07 feet:

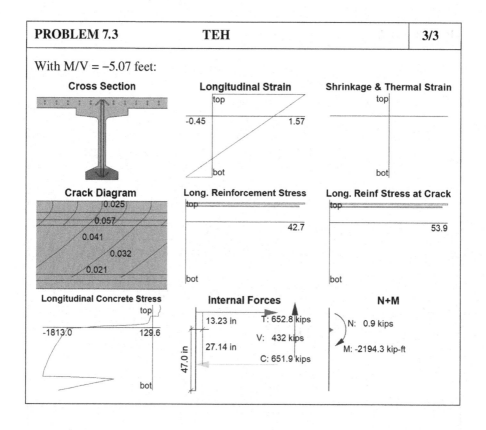

Cross Section

Longitudinal Strain

top

−0.45   1.57

bot

Shrinkage & Thermal Strain

top

bot

Crack Diagram

0.025

0.057

0.041

0.032

0.021

Long. Reinforcement Stress

top

42.7

bot

Long. Reinf Stress at Crack

top

53.9

bot

Longitudinal Concrete Stress

top

−1813.0   129.6

bot

Internal Forces

13.23 in   T: 652.8 kips

V: 432 kips

27.14 in

47.0 in

C: 651.9 kips

N+M

N: 0.9 kips

M: -2194.3 kip-ft

PROBLEM 7.4	TEH	1/1

From Problem 7-1.

$$I' = 280,175 \text{ in}^4$$

$$M_{ps} = 21,146 \text{ in-k}$$

$$W_{sw} = 0.686 \text{ ku}$$

$$\Delta_{sw} = \frac{5(0.686/12)(89.25 \times 12)^4}{384[1,820\sqrt{7}](280,175)}$$

$$\Rightarrow \Delta_{sw} = 0.726'' \downarrow$$

$$\Delta_{ps} = \frac{1}{8} ML^2/(EI')$$

$$= \frac{21,146(89.25 \times 12)^2}{8(1820\sqrt{7})(280,175)}$$

$$\Rightarrow \Delta_{ps} = 2.247'' \uparrow$$

$$\Delta_{net} = 2.247'' \uparrow + 0.726'' \downarrow$$

$$= 1.521'' \uparrow$$

PROBLEM 7.5	TEH	1/5

Compute M_{cr} using the BT-54 Girder acting composite with the deck.

Use $f'_c = 4$ ksi since the continuity diaphragm is cast with the deck.

$f_r = 0.24\sqrt{4} = 0.48$ ksi

Response 2000 Model
"CEE4380-P07.05"

For the initial trial, try 2 strands in each of the bottom 2 rows extended into the diaphragm.

$S_{bott} = 14,770$ in^3 from R2000 model

$M_{cr} = 0.48$ ksi $\times 14,770$ in$^3 = 7,090$ in-k

$= 591$ ft·k

$\Rightarrow 1.2 M_{cr} = 709$ ft·k

With the 4 strands only extended

$\Rightarrow$ when $M_{SER} = 591$ ft·k

$\Rightarrow f_{sp} = 189.6$ ksi

> 150 ksi, No Good

With 6 strands (3 in each of bottom 2 rows)

$\Rightarrow$ when $M_{SER} = 591$ ft·k

$f_{sp} = 126.4$ ksi < 150 ksi

ok

PROBLEM 7.5	TEH	2/5

With 6 extended strands.

$$M_n = 1,209 \text{ ft·k}$$

$$\phi M_n = .9 \times 1,209$$

$$= 1,088 \text{ ft·k}$$

$$> (M_u)_{STR} = 709 \text{ ft·k}$$

$$\Rightarrow \text{Strength Limit State ok}$$

Required Embedded Length of Strands.

$$\ell_{dsh} = (126.4 \times 0.228) + 8$$

$$\Rightarrow \ell_{dsh} = 36.8''$$

Bend & Extend 6 strand no less than 37'' into the continuity diaphragm at the Pier

    3 strands in bottom row
    3 strands in 2nd row

Note: in the R2000 model the default concrete tensile strength was set to a very low value to ensure tensile crack formation & a correct estimate for f_{sp}.

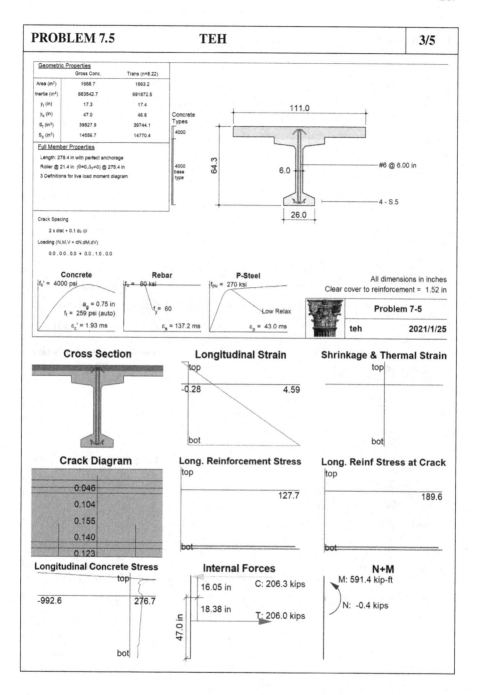

Geometric Properties

	Gross Conc.	Trans (n=8.22)
Area (in²)	1658.7	1663.2
Inertia (in⁴)	683542.7	691872.5
y_t (in)	17.3	17.4
y_b (in)	47.0	46.8
S_t (in³)	39527.9	39744.1
S_b (in³)	14556.7	14770.4

Full Member Properties

Length: 278.4 in with perfect anchorage
Roller @ 21.4 in (θ=0,Δ_y=0) @ 278.4 in
3 Definitions for live load moment diagram

Concrete Types
4000

4000 base type

Crack Spacing

2 x dist + 0.1 d_b /ρ

Loading (N,M,V + dN,dM,dV)

0.0 , 0.0 . 0.0 + 0.0 , 1.0 . 0.0

111.0

64.3

6.0   #6 @ 6.00 in

4 - S.5

26.0

Concrete
f_c' = 4000 psi
a_g = 0.75 in
f_t = 259 psi (auto)
$ε_c'$ = 1.93 ms

Rebar
f_u = 60 ksi
f_y = 60
$ε_s$ = 137.2 ms

P-Steel
f_{pu} = 270 ksi
Low Relax
$ε_p$ = 43.0 ms

All dimensions in inches
Clear cover to reinforcement = 1.52 in

Problem 7-5	
teh	2021/1/25

Cross Section

Longitudinal Strain
top
-0.28   4.59
bot

Shrinkage & Thermal Strain
top
bot

Crack Diagram
0.046
0.104
0.155
0.140
0.123

Long. Reinforcement Stress
top
127.7
bot

Long. Reinf Stress at Crack
top
189.6
bot

Longitudinal Concrete Stress
top
-992.6   276.7
bot

Internal Forces
16.05 in   C: 206.3 kips
18.38 in   T: 206.0 kips
47.0 in

N+M
M: 591.4 kip-ft
N: -0.4 kips

PROBLEM 7.5   **TEH**   **4/5**

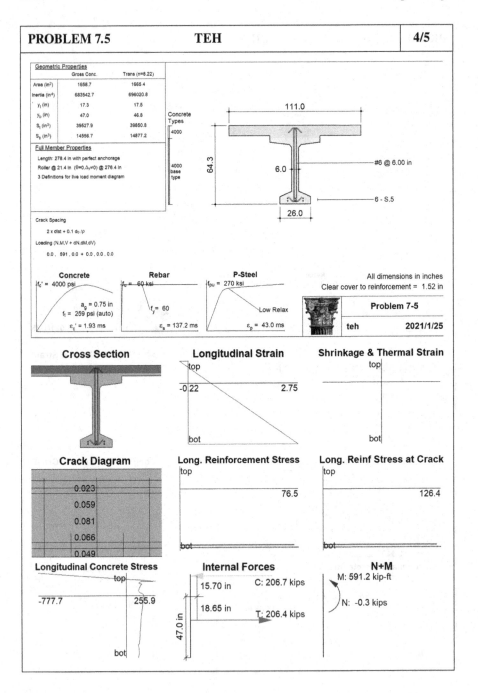

Geometric Properties

	Gross Conc.	Trans (n=8.22)
Area (in²)	1658.7	1665.4
Inertia (in⁴)	683542.7	696020.8
y_t (in)	17.3	17.5
y_b (in)	47.0	46.8
S_t (in³)	39527.9	39850.8
S_b (in³)	14556.7	14877.2

Full Member Properties

Length: 278.4 in with perfect anchorage
Roller @ 21.4 in (θ=0,Δy≠0) @ 276.4 in
3 Definitions for live load moment diagram

Concrete Types
4000
4000 base type

111.0
64.3
6.0
26.0
#6 @ 6.00 in
6 - S.5

Crack Spacing

2 x dist + 0.1 d_b /ρ

Loading (N,M,V + dN,dM,dV)

0.0 , 591 , 0.0 + 0.0 , 0.0 , 0.0

Concrete
f_c' = 4000 psi
a_g = 0.75 in
f_t = 259 psi (auto)
ε_c = 1.93 ms

Rebar
f_u = 60 ksi
f_y = 60
ε_s = 137.2 ms

P-Steel
f_{pu} = 270 ksi
Low Relax
ε_p = 43.0 ms

All dimensions in inches
Clear cover to reinforcement = 1.52 in

Problem 7-5	
teh	2021/1/25

Cross Section

Longitudinal Strain
top
-0.22   2.75
bot

Shrinkage & Thermal Strain
top
bot

Crack Diagram
0.023
0.059
0.081
0.066
0.049

Long. Reinforcement Stress
top
76.5
bot

Long. Reinf Stress at Crack
top
126.4
bot

Longitudinal Concrete Stress
top
-777.7   255.9
bot

Internal Forces
15.70 in   C: 206.7 kips
18.65 in   T: 206.4 kips
47.0 in

N+M
M: 591.2 kip-ft
N: -0.3 kips

PROBLEM 7.5	TEH	5/5

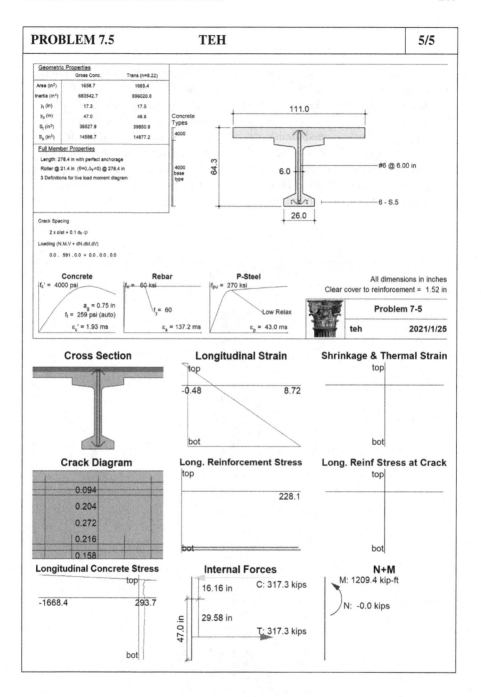

Geometric Properties

	Gross Conc.	Trans (n=8.22)
Area (in²)	1658.7	1665.4
Inertia (in⁴)	683542.7	696020.8
y_t (in)	17.3	17.5
y_b (in)	47.0	46.8
S_t (in³)	39527.9	39850.8
S_b (in³)	14556.7	14877.2

Full Member Properties

Length: 278.4 in with perfect anchorage
Roller @ 21.4 in (θ=0,Δ_y≠0) @ 278.4 in
3 Definitions for live load moment diagram

Concrete Types
4000
4000 base type

Crack Spacing

2 x dist + 0.1 d_b /p

Loading (N,M,V + dN,dM,dV)

0.0 , 591 , 0.0 + 0.0 , 0.0 , 0.0

111.0

64.3

6.0 — #6 @ 6.00 in

26.0

6 - S.5

Concrete
f_c' = 4000 psi
a_g = 0.75 in
f_t = 259 psi (auto)
ε_c' = 1.93 ms

Rebar
f_u = 60 ksi
f_y = 60
ε_s = 137.2 ms

P-Steel
f_{pu} = 270 ksi
Low Relax
ε_p = 43.0 ms

All dimensions in inches
Clear cover to reinforcement = 1.52 in

Problem 7-5	
teh	2021/1/25

Cross Section

Longitudinal Strain
top
-0.48   8.72
bot

Shrinkage & Thermal Strain
top
bot

Crack Diagram
0.094
0.204
0.272
0.216
0.158

Long. Reinforcement Stress
top
228.1
bot

Long. Reinf Stress at Crack
top
bot

Longitudinal Concrete Stress
top
-1668.4   293.7
bot

Internal Forces
16.16 in
29.58 in
47.0 in
C: 317.3 kips
T: 317.3 kips

N+M
M: 1209.4 kip-ft
N: -0.0 kips

7.10   EXERCISES

E7.1.

An interior BT-72 girder has forty ½-inch-diameter, 270K low-lax strands arranged as shown in Figure E7.1. Girder concrete strength f'_c = 7,000 psi. Composite deck thickness is 8.25 inches with a girder spacing S of 11 ft 6 inches. Deck concrete f'_c = 4,000 psi. The initial pull on the strands is 75% of f_{pu} and the estimated total losses are 23.55 ksi. Young's modulus for the strands is E = 28,500 ksi. Determine the flexural resistance, ϕM_n, in positive bending of the composite girder. Haunch depth is 2 inches. All template rows shown are 2 inches above the previous row. The bottom row is 2.5 inches from the bottom of the girder.

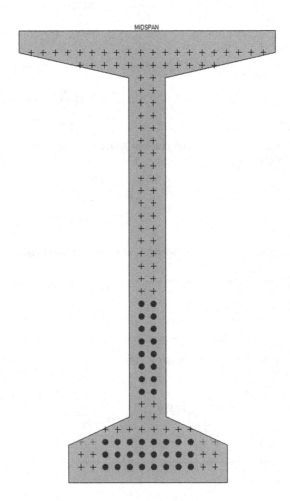

FIGURE E7.1   Exercise E7.1.

Note: Some engineers do not permit the placement of two strands per row in a web only 6 inches wide. Such practice, while resulting in a relatively congested geometry, has been used successfully. Consult local fabricators and owners.

E7.2.

The BT-63 girder shown in Figure E7.2 is made from concrete with f'_{ci} = 8 ksi and f'_c = 9.5 ksi. Strands are 0.60 inch diameter, 270K low-lax, straight strands with no de-bonding. Initial pull is 75% of f_{pu}. The girder is made composite with the 4 ksi concrete deck, which is 8.25 inches thick. Haunch thickness is 2 inches. The bottom row of strands is 2.5 inches from the bottom of the girder. Subsequent rows are 2 inches above the previous row. Initial losses due to elastic shortening are 12.7 ksi. Total final losses are estimated to be 22.3 ksi.

Strength limit state moment is -4,463 ft-kips with a simultaneous shear equal to 197 kips. Stirrups (not shown, for clarity) are single #6 spaced at 6 inches on center extending into the deck. The top mat of 16 #7 bars is located 3.7 inches below the top of the deck. The bottom mat of 12 #7 bars is located 2.1 inches from the bottom of the deck. Grade 60 reinforcing is used.

Assess the adequacy of the stirrups and deck longitudinal reinforcement to carry the required Strength limit state loads. Strands are ineffective at the girder end and are to be ignored.

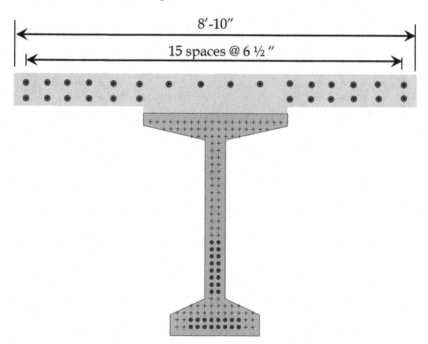

FIGURE E7.2   Exercise E7.2.

Note: Some engineers do not permit the placement of two strands per row in a web only 6 inches wide. Such practice, while resulting in a relatively congested geometry, has been used successfully. Consult local fabricators and owners.

E7.3.

The BT-63 girder shown in Figure E7.3 is made from concrete with $f'_{ci} =$ 8 ksi and $f'_c = 9.5$ ksi. Strands are 0.60 inch diameter, 270K low-lax, straight strands with no de-bonding. Initial pull is 75% of f_{pu}. The girder is made composite with the 4 ksi concrete deck, which is 8.25 inches thick. Haunch thickness is 2 inches. The bottom row of strands is 2.5 inches from the bottom of the girder. Subsequent rows are 2 inches above the previous row. Initial losses due to elastic shortening are 12.7 ksi. Total final losses are estimated to be 22.3 ksi.

Service I limit state moment is −2,720 ft-kips with simultaneous shear of 175 kips. Stirrups (not shown, for clarity) are single #6 spaced at 6 inches on center extending into the deck. The top mat of 16 #7 bars is located 3.7 inches below the top of the deck. The bottom mat of 12 #7 bars is located 2.1 inches from the bottom of the deck. Grade 60 reinforcing is used. For the deck, $f'_c = 4,000$ psi.

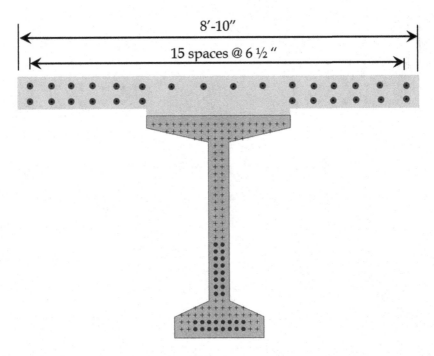

FIGURE E7.3   Exercise E7.3.

Assess the adequacy of the deck reinforcing to meet crack control requirements at the Service limit state. Strands are ineffective at the girder end and are to be ignored.

Note: Some engineers do not permit the placement of two strands per row in a web only 6 inches wide. Such practice, while resulting in a relatively congested geometry, has been used successfully. Consult local fabricators and owners.

E7.4.

The BT-63 girder shown in Figure E7.4 is made from concrete with f'_{ci} = 8 ksi and f'_c = 9.5 ksi. Strands are 0.60 inch diameter, 270K low-lax, straight strands with no de-bonding. Initial pull is 75% of f_{pu}. The girder is made composite with the 4 ksi concrete deck, which is 8.25 inches thick. Haunch thickness is 2 inches. The bottom row of strands is 2.5 inches from the bottom of the girder. Subsequent rows are 2 inches above the previous row. Initial losses due to elastic shortening are 12.7 ksi. Total final losses are estimated to be 22.3 ksi.

Strength limit state moment, M_u, is +7,208 ft-kips at a distance 40 percent of the span length from the abutment (the point of maximum positive moment). The shear, V_u, acting simultaneously with the moment, is 46 kips. Grade 60 reinforcing is used. Stirrups (not shown, for clarity) are single #6 at 18 inches.

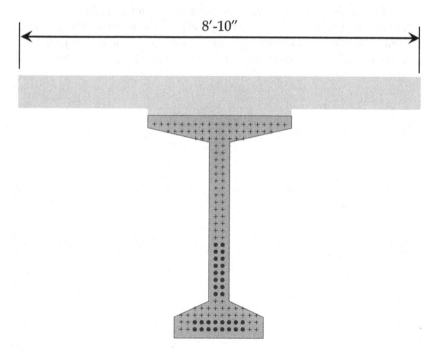

FIGURE E7.4   Exercise E7.4.

Determine the design flexural and shear resistances, ϕM_n and ϕV_n, and compare with the required values, M_u and V_u.

Note: Some engineers do not permit the placement of twi strands per row in a web only 6 inches wide. Such practice, while resulting in a relatively congested geometry, has been used successfully. Consult local fabricators and owners.

E7.5.

For the girder of exercise E7.4, estimate midspan stresses and deflections at release. Compare the stresses to permissible values. The girder length is 113 ft.

E7.6.

A 125-ft-long BT-72 bridge girder has forty 0.60-inch-diameter, 270K low-lax strands, 14 of which are draped at 0.4L. The strand group is raised a total of 4 ft 4 inches from the drape point to the girder end. Initial pull is 75 percent of f_{pu}. Determine (a) the required hold-down force at the drape point for fabrication of the girder and (b) the stress in the straight portion of the draped strands under initial pull.

E7.7.

For the girder of exercise E7.4, determine the required non-prestressed mild reinforcement in the top of the girder to resist tension prior to losses.

E7.8.

For the girder in exercise E7.3, determine the required number and length of bent strands at the girder ends to provide the required positive moment connection. Assume that the girder age at continuity is no less than 90 days and thus, restraint moments may be ignored in the design.

8 Bridge Girder Bearings

Two popular bearing configurations are (a) elastomeric bearing pads, plain or reinforced with steel layers, and (b) steel assemblies, as either fixed or expansion bearings. Elastomeric pads may also be constructed to behave as either fixed or expansion bearings.

Each of these bearing types, with modification, may be used as seismic isolation bearings as well.

Other bearing types include metal roller systems, PTFE sliding surface systems, pot bearings, and disc bearings. These are not covered here, and Chapter 14 of the AASHTO LRFD BDS contains detailed requirements for such.

8.1 ELASTOMERIC BEARINGS

This section summarizes design criteria for elastomeric bearings in the AASHTO LRFD BDS. The reader is referred to Section 14.7.5 of the AASHTO LRFD BDS for the basis of the material presented here, and for further explanations and discussion on the design of elastomeric bearings for bridge girders.

Figure 8.1 depicts details for an expansion elastomeric bearing used in a bridge in Clay County, Tennessee on State Route 52.

Prior to discussion of design criteria, a definition of terms will be helpful.

- h_{rt} = total elastomer thickness, exclusive of steel layers
- h_{ri} = thickness of ith elastomer layer
- G = specified shear modulus of elastomer, ksi
- Δ_S = service shear deformation
- S_i = shape factor
- L = plan dimension of a rectangular bearing perpendicular to the axis of rotation
- W = plan dimension of a rectangular bearing parallel to the axis of rotation
- D = diameter of a round bearing
- d = diameter of a hole through an elastomeric bearing
- D_r = dimensionless rotation coefficient (0.500 for rectangular bearings, 0.375 for circular bearings)
- D_a = dimensionless axial coefficient (1.4 for rectangular bearings, 1.0 for circular bearings)
- γ_a = shear strain caused by axial load
- γ_r = shear strain caused by rotation
- γ_s = shear strain caused by shear displacement
- $\gamma_{a,st}$ = shear strain due to non-traffic axial load

DOI: 10.1201/9781003265467-8

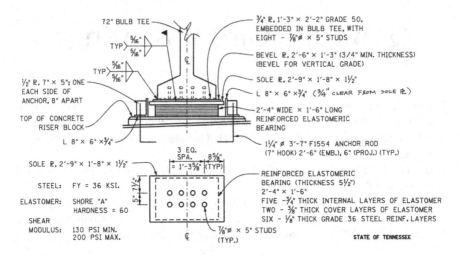

FIGURE 8.1   State Route 52 over branch – elastomeric bearing details.

- $\gamma_{a,cy}$ = shear strain due to traffic-induced axial load
- $\gamma_{r,st}$ = shear strain due to non-traffic rotation
- $\gamma_{r,cy}$ = shear strain due to traffic-induced rotation
- $\gamma_{s,st}$ = shear strain due to non-traffic shear
- $\gamma_{s,cy}$ = shear strain due to traffic-induced shear
- σ_s = Service limit state total compressive stress on the bearing (all load factors = 1.00)
- σ_L = Service limit state live load compressive stress on the bearing (load factor = 1.00)
- σ_{hyd} = peak hydrostatic stress
- θ_s = service limit state rotation; to include an additional 0.005 radians for miscellaneous uncertainties
- h_s = thickness of steel reinforcement layers
- n = number of interior layers of elastomer; when the thickness of the exterior layer of elastomer is equal to or greater than one-half the thickness of an interior layer, the parameter, n, may be increased by one-half for each such exterior layer.
- F_y = yield stress of steel layers
- ΔF_{TH} = Category 'A' fatigue threshold (24 ksi)

The cover layer thickness of a reinforced elastomeric bearing is to be no more than 70% of the internal layer thickness. All internal elastomer layers shall be the same thickness.

The shape factor, S_i, of an elastomer layer is defined as the plan area of the layer divided by the area of perimeter free to bulge. The shape factor is defined in Equation 8.1 for rectangular bearings, or in Equation 8.2 for round bearings. Although the equations do allow for the presence of holes in the elastomeric bearing to accommodate anchor rods, such practice is generally to be discouraged. It is usually possible to place the anchor rods outside the limits of the elastomeric bearing.

$$S_i = \frac{LW - \sum \frac{\pi d^2}{4}}{h_{ri}\left[2L + 2W + \sum \pi d\right]}, \text{ rectangular} \tag{8.1}$$

$$S_i = \frac{D^2 - \sum d^2}{4h_{ri}\left(D + \sum d\right)}, \text{ round} \tag{8.2}$$

In addition to plan dimensions and elastomer and steel layer details, elastomeric bearing definition and design must include a specified shear modulus, G. AASHTO requires that (a) the shear modulus specified on the plans must be in the range of 0.080 to 0.175 ksi, (b) a tolerance of $\pm 15\%$ on the specified shear modulus is to be incorporated into the design, and (c) the least favorable value is to be assigned to the shear modulus used in calculations. Generally speaking, the minimal G (0.85 times the specified value) within the range will be the least favorable for stress checks and the maximal G (1.15 times the specified value) within the range will be the least favorable in rotation capacity calculations or in determining the loads delivered to the substructures by the bearings.

Not only a shear modulus, G, but also a minimum low-temperature grade, should be specified for elastomeric material used in reinforced elastomeric bearings. Elastomeric materials stiffen as temperature decreases. In Section 14.7.5.2 of the AASHTO LRFD BDS, five temperature zones are mapped, with Zone A being the least severe and Zone E the most severe. The minimum low-temperature grade of elastomer suitable for each zone, with forces delivered to substructures as presented later, is summarized below. It is permissible to use a lower-than-recommended grade, but with increased forces delivered to the substructures for design. Refer to Section 14.7.5.2 of the AASHTO LRFD BDS for such provisions.

- Low-temperature Zone A, minimum low-temperature grade $= 0$
- Low-temperature Zone B, minimum low-temperature grade $= 2$
- Low-temperature Zone C, minimum low-temperature grade $= 3$
- Low-temperature Zone D, minimum low-temperature grade $= 4$
- Low-temperature Zone E, minimum low-temperature grade $= 5$

For thermal expansion and contraction, the design movement is taken to be equal to 65% of the total thermal movement range to account for variation in installation temperatures.

Expansion bearings are subject to shear displacement, whereas fixed bearings are not. For expansion bearings, the Service limit state shear deformation is limited to that given by Equation 8.3.

$$\Delta_s \leq \frac{h_{rt}}{2} \tag{8.3}$$

The minimum thickness for steel layers is 0.0625 inches. Steel layers must satisfy requirements at both the Service and Fatigue limit states, as defined by Equations 8.4 and 8.5, respectively.

$$h_s \geq \frac{3h_{ri}\sigma_s}{F_y} \tag{8.4}$$

$$h_s \geq \frac{2h_{ri}\sigma_L}{\Delta F_{TH}} \tag{8.5}$$

Stability checks for elastomeric bearings are given in Equation 8.6, for expansion bearings, and Equation 8.7 for fixed bearings. Stability checks for circular bearings are to be made using $L = W = 0.8D$. If $2A \leq B$, then the bearing is stable and the stress checks given by the Equations need not be made. Otherwise, the stress check must be satisfied to ensure stability of the bearing.

$$\sigma_s \leq \frac{GS_i}{2A - B} \tag{8.6}$$

$$\sigma_s \leq \frac{GS_i}{A - B} \tag{8.7}$$

$$A = \frac{1.92\dfrac{h_{rt}}{L}}{\sqrt{1 + 2\dfrac{L}{W}}} \tag{8.8}$$

$$B = \frac{2.67}{\left(S_i + 2.0\right)\left(1 + \dfrac{L}{4W}\right)} \tag{8.9}$$

Shear strains for bearing design are determined using Equations 8.10 through 8.13, with the limiting values provided in Equations 8.14 and 8.15.

$$\gamma_a = D_a \frac{\sigma_s}{GS_i} \tag{8.10}$$

$$\gamma_r = D_r \left(\frac{L}{h_{ri}}\right)^2 \frac{\theta_s}{n}, \text{rectangular} \tag{8.11}$$

$$\gamma_r = D_r \left(\frac{D}{h_{ri}}\right)^2 \frac{\theta_s}{n}, \text{circular} \tag{8.12}$$

$$\gamma_s = \frac{\Delta_s}{h_{rt}} \tag{8.13}$$

$$\gamma_{a,st} \leq 3.0 \tag{8.14}$$

$$\left(\gamma_{a,st} + \gamma_{r,st} + \gamma_{s,st}\right) + 1.75\left(\gamma_{a,cy} + \gamma_{r,cy} + \gamma_{s,cy}\right) \leq 5.0 \tag{8.15}$$

For bearings with bonded plates on both top and bottom, the additional requirements given in Equations 8.16 through 8.22 must be satisfied. The coefficient, B_a, is taken to be equal to 1.6 unless a more refined solution can be justified. For bearings without externally bonded plates, a restraint system to prevent lateral movement is required whenever Equation 8.23 is found to be valid.

$$\sigma_{hyd} = 3GS_i^3 \frac{\theta_{s2}}{n} C_\alpha \tag{8.16}$$

$$C_\alpha = \frac{4}{3}\left[\left(\alpha^2 + \frac{1}{3}\right)^{1.5} - \alpha\left(1 - \alpha^2\right)\right] \tag{8.17}$$

$$\alpha = \frac{\varepsilon_a}{S_i} \cdot \frac{n}{\theta_{s2}} \tag{8.18}$$

$$\varepsilon_a = \frac{\sigma_{s2}}{3B_a GS_i^2} \tag{8.19}$$

$$\sigma_{hyd} \leq 2.25G \tag{8.20}$$

$$\sigma_{s2} = \frac{P_D + 1.75P_L}{A} \tag{8.21}$$

$$\theta_{s2} = \theta_D + 1.75\theta_L \tag{8.22}$$

$$\frac{\theta_s}{n} \geq \frac{3\varepsilon_a}{S_i} \tag{8.23}$$

The elastomeric bearing, once final design dimensions and properties have been established to satisfy rotational, strain, stress, and stability checks, must be analyzed to determine the magnitude of forces delivered to the substructure by the bearing. Equations 8.24 and 8.25 provide a means of determining both the shear force and the moment delivered to the supporting member by the bearing.

$$H_{bu} = GA\frac{\Delta_u}{h_{rt}}$$
(8.24)

$$M_u = 1.60\left(0.5E_cI\right)\frac{\theta_s}{h_{rt}}$$
(8.25)

$$E_c = 4.8GS_i^2$$
(8.26)

$$I = \frac{WL^3}{12}, \text{rectangular}$$
(8.27)

$$I = \frac{\pi D^4}{64}, \text{circular}$$
(8.28)

8.2   STEEL ASSEMBLY BEARINGS

For large vertical reactions and large thermal movement requirements, it may be necessary to use steel assembly bearings. Figure 8.2 shows an elevation at the girder end for the expansion bearing used for a bridge over Center Hill Lake in Smithville Tennessee on State Route 26. Figure 8.3 shows the fixed counterpart for the Center Hill Lake Bridge. Fixed bearings were used at the west abutment and at the four piers for the 1,545-foot-long steel girder bridge. Expansion bearings were used for the girders at the east abutment, where the entire thermal expansion and contraction requirements were accommodated.

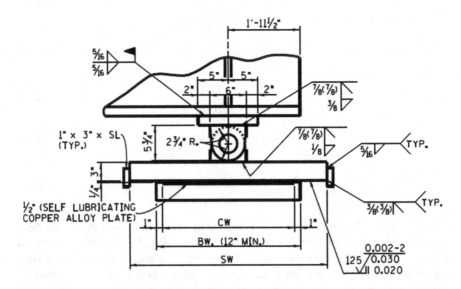

FIGURE 8.2 Expansion bearing for State Route 26 over Center Hill Lake, Smithville, Tennessee.

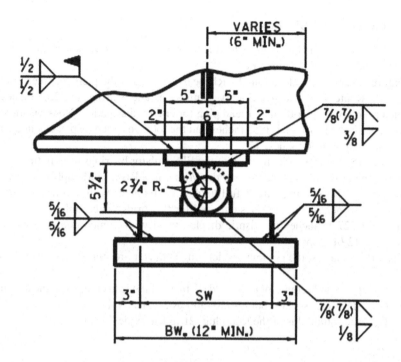

FIGURE 8.3   Fixed bearing for State Route 26 over Center Hill Lake, Smithville, Tennessee.

Observe, from Figure 8.2, that the total movement capacity of the bearing is equal to $SW-BW$, before the restrainers engage.

Either bronze or copper slider plates, at the contractor's option, were permitted in the contract documents for the bearings shown in Figure 8.2. The AASHTO LRFD BDS, in Section 14.7.7.2, specifies a coefficient of friction equal to 0.10 for self-lubricating bronze and 0.40 for other slider plate materials. Frictional forces (FR) acting laterally for either expansion or contraction must be accommodated in the substructure design. Such forces would need to be based on the frictional forces developed with the higher coefficient.

Bearing stress at the Service limit state is limited to 2.0 ksi for Type 1 and 2 bronze plates in the AASHTO LRFD BDS, Section 14.7.7.3. Bearing stress on concrete at the Strength limit state is limited to that given by Equation 8.29, where A_1 is the plan area of the bearing contact surface. See Section 5.6.5 of the AASHTO LRFD BDS. Bearing pressure checks on both the bronze/copper slider plate and on the concrete support should be made under the fully extended condition for expansion bearings. The coefficient, m, may conservatively be taken to be equal to 1.0. When the supporting concrete is wider than the bearing on all sides, m may reach values as high as 2.0. Refer to Section 5.6.5 of the AASHTO LRFD BDS for methods which may be used to justify m-values larger than 1.0.

$$\phi P_n = 0.70\left[0.85 f_c' A_1 m\right] \qquad (8.29)$$

8.3 ISOLATION BEARINGS

Seismic isolation bearings include lead-rubber bearings (LRB) and friction pendulum systems (FPS), among others.

Figure 8.4 shows an elevation of LRB bearings used at the piers for Interstate 40 over State Route 5 in Madison County, Tennessee. The bearings are 22.5 inches in diameter with 3.75-inch-diameter lead plug for energy dissipation. Non-seismic displacement for the bearings is 0.52 inches and the total designed displacement during a seismic event is 8.30 inches.

Figure 8.5 is a schematic diagram of the FPS isolator bearings used in the retrofit of the main piers for Interstate 40 over the Mississippi River in Memphis, Tennessee (Figure 8.6). The bearings are 8 ft 10inches in diameter with a concave surface radius of 244 inches. The FPS system used for this bridge is capable of accommodating up to 27.25 inches of seismic displacement with a maximum vertical load capacity of 12,611 kips.

For LRB isolators, the parameters defining behavior are summarized as follows:

- f factor to account for post-yield stiffness of the lead core (1.1, typical values)
- G elastomer shear modulus (50–300 psi, typical values)
- G_p shear modulus of the lead plug (21.75 ksi, typical values)

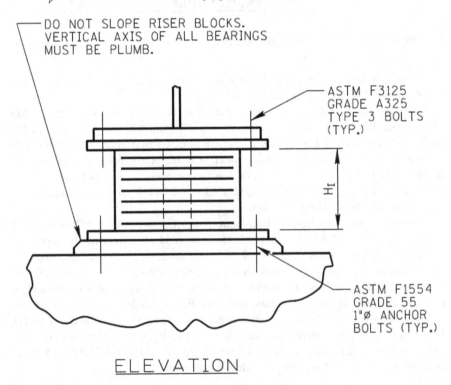

ELEVATION

FIGURE 8.4   Interstate 40 over State Route 5 isolation bearings.

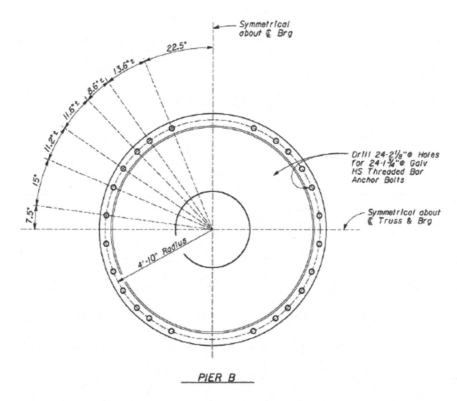

FIGURE 8.5 Friction pendulum system (FPS) bearings for the Hernando de Soto Bridge in Memphis, Tennessee.

- T_r total rubber (elastomer) thickness
- A_b bonded area of rubber
- d_b diameter of circular bearing
- d_L diameter of lead plug ($d_b/6$ to $d_b/3$, typical values)
- f_{yL} yield stress in shear for lead (1.3–1.5 ksi, typical values)
- ψ stress modifier for lead plug:
  - 1 for EQ-load
  - 2 for wind/braking
  - 3 for thermal expansion
- α post-yield stiffness ratio, typical values:
  - 0.10 for EQ
  - 0.125 for wind/braking
  - 0.20 for thermal expansion
- γ_c shear strain due to compression
- γ_r shear strain due to rotation
- $\gamma_{s,s}$ shear strain due to non-seismic displacement
- $\gamma_{s,eq}$ shear strain due to seismic displacement
- t_i thickness of an individual elastomer layer

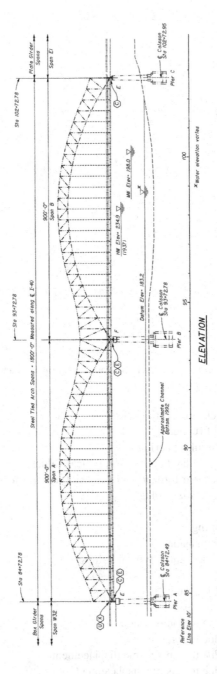

FIGURE 8.6 Hernando de Soto Bridge in Memphis, Tennessee.

- θ bearing rotation (include a 0.005 radian contingency)
- B bonded plan dimension in the direction of loading (d_b for circular)
- ξ_{EFF} effective damping imparted into a system through inelastic behavior
- B_L response modification divisor to account for added damping
- Q_d characteristic strength
- k_d post-yield stiffness
- Δ_s lateral non-seismic displacement demand on an isolation bearing
- D_{ISO} lateral seismic displacement demand on an isolation bearing
- S shape factor; bonded plan area divided by the side area free to bulge
- D_r shear strain factor
  - 0.375, circular bearing
  - 0.500, rectangular bearing
- D_C shape coefficient
  - 1.000, circular bearing
  - 1.000, rectangular bearing
- σ_s compressive stress $= P/A_b$

Design criteria for LRB isolators used in bridge structures may be found in AASHTO GS ISO design specifications (AASHTO, 2014), supplemented with literature on the subject (Stafford et al., 2008; Warn, 2002). A subset of applicable criteria is summarized in Equations 8.30 through 8.46. Loads that contribute to the static components of deformation include wind, dead load, and thermal effects. Loads that are assumed to contribute to the cyclic components of deformation include live load, braking forces, and seismic effects.

Equations for these systems are typically approximate and final properties are generally established through rigorous testing of the bearings.

$$A_b = \frac{\pi\left(d_b^2 - d_L^2\right)}{4} \tag{8.30}$$

$$k_d = f\,\frac{GA_b}{T_r} \tag{8.31}$$

$$k_i = \frac{G_p A_p + GA_b}{T_r} \tag{8.32}$$

$$F_y = \frac{1}{\psi}\cdot f_{yL}\cdot\frac{\pi d_L^2}{4} \tag{8.33}$$

$$S = \frac{d_b^2 - d_L^2}{4 d_b t_i} \tag{8.34}$$

$$\gamma_c = \frac{D_c \sigma_s}{GS} \tag{8.35}$$

$$\gamma_r = \frac{D_r B^2 \theta}{t_i T_r} \tag{8.36}$$

$$\gamma_{s,s} = \frac{\Delta_s}{T_r} \tag{8.37}$$

$$\gamma_{s,eq} = \frac{D_{ISO}}{T_r} \tag{8.38}$$

$$\left(\gamma_c + \gamma_r + \gamma_{s,s}\right)_{static} + 1.75\left(\gamma_c + \gamma_r + \gamma_{s,s}\right)_{cyclic} \le 5.0 \tag{8.39}$$

$$\left(\gamma_c\right)_{static} \le 3.0 \tag{8.40}$$

$$\left(\gamma_c + 0.50\gamma_r + \gamma_{s,s} + \gamma_{s,eq}\right)_{total} \le 5.5 \tag{8.41}$$

$$K_{EFF} = k_d + \frac{Q_d}{D_{ISO}} \tag{8.42}$$

$$\xi_{EFF} = \frac{2Q_d\left(D_{ISO} - D_y\right)}{\pi\left(D_{ISO}\right)^2 K_{EFF}} \tag{8.43}$$

$$D_y = \frac{Q_d}{k_d} \cdot \frac{\alpha}{1-\alpha} \tag{8.44}$$

$$F_y = \frac{Q_d}{1-\alpha} \tag{8.45}$$

$$B_L = \left(\frac{\xi_{EFF}}{0.05}\right)^{0.30} \tag{8.46}$$

Design parameters for friction pendulum systems (FPS) include:

- μ dynamic friction coefficient (0.03–0.12, typical values)
- R radius of concave surface
- W vertical load
- D_{vert} vertical displacement due to concave sliding surface

A summary of relationships for FPS isolation systems is provided in Equations 8.47 through 8.53.

$$Q_d = \mu W \tag{8.47}$$

$$k_d = \frac{W}{R} \tag{8.48}$$

$$k_{eff} = \frac{W}{R} + \mu \cdot \frac{W}{D_{ISO}} \tag{8.49}$$

$$\xi_{eff} = \frac{2}{\pi} \cdot \frac{\mu}{\mu + D_{ISO}/R} \tag{8.50}$$

$$\varphi = \sin^{-1} \frac{D_{ISO}}{R} \tag{8.51}$$

$$D_{vert} = R\left(1 - \cos\varphi\right) \cong \frac{D_{ISO}^2}{2R} \tag{8.52}$$

$$\mu \le \frac{D_{ISO}}{R} \le 0.15 \tag{8.53}$$

Although not a specification requirement, limits on D_{ISO}/R given in Equation 8.53 are generally considered to be good practice. The lower limit establishes improved re-centering capability. The upper limit assures that the small rotation angle assumption about the center of curvature of the concave surface is valid. As with LRB devices, the equations for FPS systems are approximate and final properties are established through testing.

8.4 ANCHOR RODS

The design shear resistance, ϕR_n, for ASTM F1554 anchor rods in shear is given by Equation 8.54, where F_{ub} is the specified minimum tensile strength of the anchor rod, A_b is the nominal area of the anchor rod, and N_s is the number of shear planes, typically equal to 1 for anchored bearings. Refer to Section 6.13.2.12 of the LRFD BDS for additional information. Chapter 1 of this text includes a summary of the mechanical properties and available diameters for ASTM F 1554 anchor rods.

$$\phi P_n = 0.75 \left[0.50 F_{ub} A_b N_s \right] \tag{8.54}$$

Anchor rods are commonly used to carry lateral forces from the bearings to the pier cap in both steel girder and concrete girder bridges. These anchors may be located at the bearings, embedded in pier diaphragms, or both. The number of shear planes, N_s, for such anchors is one.

8.5 SOLVED PROBLEMS

Problem 8.1

A prestressed concrete beam bridge incorporates the bearing design shown in the end elevation (Figure P8.1). Dead load rotation may be taken as zero.

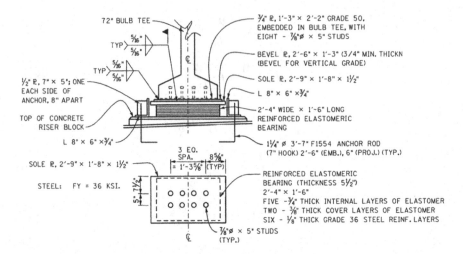

FIGURE P8.1   Problem P8.1.

Live load rotation is 0.009 radians. Miscellaneous static rotation is 0.005 radians. Axial loads are 425 kips (DL) and 275 kips (LL). The anticipated thermal movement is 2.00 inches each way. The specified shear modulus is 175 psi. AASHTO requires that a ± 15% deviation in specified shear modulus be incorporated into all design checks, using the least favorable value in the range.

Check the bearing for the following AASHTO requirements:

a) shear deformation limits
b) strain limits
c) stability requirements
d) steel reinforcement requirements
e) dimensional requirements
f) determination of the shear and moment the bearing delivers to the substructure

Problem 8.2

For the Sligo bridge expansion bearings, determine the required dimensions *SW* and *BW*. Establish the maximum design resistance for the bearing reaction based (a) on the slider plate and (b) on the concrete cap. The bridge is a steel girder bridge in a moderate climate. The entire expansion requirement for the 1,545-ft-long bridge is to be taken by expansion bearings at Abutment 2. Refer to Figures 8.2 and 8.3.

The girder reactions are:

$$R_{DC} = 236 \text{ kips} \quad R_{DW} = 48 \text{ kips} \quad R_{LL+IM} = 167 \text{ kips}$$

The baseplate area is $BL \times BW$. The slider plate area is $CW \times L$. Take:

- $CW = BW{-}2$
- $BL = 36$ inches
- $CL = 28$ inches

Problem 8.3

An anchor rod made from ASTM F 1554 material is to be used to carry seismic shear forces from the superstructure to the intermediate pier for a two-span bridge. Span lengths are 140 ft each and the bridge is 45 ft wide. Five girders are spaced 9 ft 6 inches apart on center. The total superstructure weight is 13.1 kips per foot, including two lanes of HL-93 uniform lane loading. The anchors are embedded in a 1-ft 6-inch-wide diaphragm at the pier. At the project site, $S_{DS} = 0.854$ g and $S_{DI} = 0.733$ g. The fundamental period of the structure is 0.95 seconds. For preliminary design, a simplified analysis is to be used for sizing the anchors. The elastic seismic force is to be reduced by a factor of 1.5 for the simplified analysis. Final design will ensure that the anchors can resist the overstrength plastic shear of the pier. Design the preliminary anchor configuration (by determining the anchor grade, the anchor diameter, and the number of required anchors at the pier).

Problem 8.4

A 31.5-inch-diameter isolation bearing (LRB) consists of 15 internal elastomer layers, each 0.75 inches thick. External layers are each 0.375 inches thick and the internal steel layers (16) are each 0.125 inches thick. The elastomer shear modulus is $G = 55$ psi. Shear yield stress for the 8-inch-diameter lead plug is 1.3 ksi. The shear modulus for the lead plug is $G = 21{,}750$ psi. The maximum seismic displacement is estimated to be 20 inches. Service level rotation is 0.010 radians (100% cyclic) and deflections due to non-seismic loads are 3.8 inches (TU) and 0.7 inches (BR). Determine:

- the initial stiffness and yielded stiffness for the bearing
- the yield force and yield displacement for the bearing
- the effective damping at maximum seismic displacement
- the force delivered to the substructure by the bearing at maximum seismic displacement.
- The maximum compressive force to be permitted on the bearing if the ratio of total dead load to total live load (D/L) is 1.68).

PROBLEM 8.1	TEH	1/3

a) Shear deformation

$$h_{rt} = 5(.75) + 2(.375) = 4.50''$$

$$2\Delta_s = 2 \times 2.00 = 4.00''$$

$$h_{rt} > 2\Delta_s \implies ok$$

b) Strain limits

$$D_a = 1.4 \quad , \quad D_r = 0.500$$

$$\sigma_s = \frac{425 + 275}{28 \times 18} = 1.389 \text{ ksi}$$

$$S_i = \frac{28 \times 18}{2(.75)(28 + 18)} = 7.304$$

Use $G_{MIN} = .85 \times 175 = 149$ psi
for calculating actual strain.

$$\gamma_a = 1.40 \left(\frac{1,389}{149 \times 7.304} \right) = 1.787 \text{ rad}$$

$$\gamma_r = 0.500 \left(\frac{18}{.75} \right)^2 \frac{(.009 + .005)}{6}$$

(since $h_{ro} = 0.375 > h_{ri} = 0.75$:

$$n = 5 + \frac{1}{2} + \frac{1}{2} = 6 \text{)}$$

$$\gamma_r = 0.672 \text{ rad}$$

$$\gamma_s = \frac{2.00}{4.50} = 0.444 \text{ rad}$$

$$\gamma_{a,st} = (425/700) \times 1.787 = 1.085$$

$$\gamma_{a,cy} = (275/700) \times 1.787 = 0.702$$

PROBLEM 8.1	TEH	2/3

$$\gamma_{r,st} = \left(\frac{.005}{.014}\right) \times 0.672 = 0.240$$

$$\gamma_{r,cy} = \left(\frac{.009}{.014}\right) \times 0.672 = 0.432$$

$$\gamma_{a,st} = 1.085 < 3.0 , \text{ ok}$$

$$\gamma_{st} + 1.75 \gamma_{cy} = 1.085 + 0.240 + 0.444$$
$$+ 1.75(0.702 + 0.432)$$

$$\Rightarrow \quad = 3.753 < 5.0 , \text{ ok}$$

Strain Limits ok

c) Stability req'ts.

$$A = \frac{1.92 (4.50/18)}{\sqrt{1 + 2(18/28)}}$$

$$\Rightarrow A = 0.3175$$

$$B = \frac{2.67}{9.304 \left(1 + \frac{18}{4 \times 28}\right)} = 0.2472$$

$$2A - B = 0.3877$$

$$\frac{GS}{2A - B} = \frac{.149 \times 7.304}{0.3877} = 2.807 \text{ ksi}$$

$$\sigma_s = 1.389 \text{ ksi} < 2.807 \text{ ksi}$$

$$\Rightarrow \text{ Stability ok}$$

PROBLEM 8.1	TEH	3/3

d) Steel reinforcement.

$$\text{Service}: h_s \geqslant \frac{3(0.75)(1.389)}{36}$$

$$= 0.087''$$

$$\text{Fatigue}: h_s \geqslant \frac{2(0.75)(1.389 \times 275/100)}{24}$$

$$= 0.034''$$

$$h_s = 0.125'' > 0.087'', \text{ ok}$$

e) Dimensional req'ts.

$$h_{ro} = 3/8'' = \frac{1}{2} h_i < .70 h_i$$

$$\Rightarrow \text{ cover thickness ok}$$

f) Forces delivered to substructure.

$$\text{Use } G_{max} = 175 \times 1.15 = 201 \text{ psi}$$

$$H_u = 0.201 (28 \times 18)(2/4.50)$$

$$\Rightarrow \underline{H_u = 45.0 \text{ kips per brg.}}$$

$$I = 28(18)^3/12 = 13,608 \text{ in}^4$$

$$E_c = 4.8(.201)(7.304)^2 = 51.47 \text{ ksi}$$

$$\theta = .005 + 1.75 \times .009 = 0.02075$$

$$M_u = 1.6(.5)(51.47)(13,608)(0.02075/4.50)$$

$$\Rightarrow \underline{M_u = 2,584 \text{ in-k} = 215 \text{ ft·k}}$$
$$\underline{\text{per bearing.}}$$

PROBLEM 8.2	TEH	1/2

$\delta T = 120 - 0 = 120°F$

$\alpha = 0.0000065$

$\Delta_{Tu} = (120 \times 0.0000065)(1,545)(12)$

$\qquad = 14.46''$

$\qquad \underline{\times 1.20}$

$\qquad 17.35''$

Design for $18''$ total $= 9''$ each way.

$SW - BW = 18''$

Based on Concrete : (Strength Limit State)

$\qquad R_u = 1.25 \times 236 + 1.5 \times 48 + 1.75 \times 167$

$\qquad \Rightarrow R_u = 660 \text{ KIPS}$

$\qquad \phi R_n = 0.7 \times 0.85 \times 4 (BW \times 36)$

$\qquad = 85.68 \, BW \geqslant 660$

$\qquad \Rightarrow BW \geqslant 7.70 \text{ inches}$

However, this is an expansion bearing
& the pressure must be satisfied in the
fully extended position. With full
expansion, the reaction centerline is
$9''$ from the plate centerline.

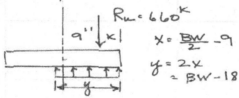

$R_u = 660^k$

$X = \dfrac{BW}{2} - 9$

$y = 2x$
$\quad = BW - 18$

PROBLEM 8.2	TEH	2/2

$\Rightarrow BW - 18 \geqslant 7.70$

$\Rightarrow BW \geqslant 25.7''$

Based on the Slider Plate: (Service)

$R_u = 236 + 48 + 167 = 451 \text{ KIPS}$

(also check in fully extended condition)

$X = \dfrac{CW}{2} - 9$

$y = 2X = CW - 18$

$(CW - 18)(CL)(2.0 \text{ ksi}) \geqslant 451^K$

$(CW - 18)(28)(2.0) \geqslant 451$

$CW - 18 \geqslant 8.05''$

$CW \geqslant 26.05''$

Take $BW = 26''$ ($> 25.7''$ req'd)

$SW = 26 + 18 = 44''$

But, based on CW, $BW = CW + 2$

Take $CW = 27'' > 26.05''$ req'd

$BW = 29'' > 25.7''$ req'd

$SW = 29 + 18 = 47''$

PROBLEM 8.3	TEH	1/1

$S_{DS} = 0.854 \qquad S_{D1} = 0.733$

$T_S = 0.733 / 0.854 = 0.858 \text{ SEC}$

$T = 0.95 \text{ SEC} > T_S$

$\Rightarrow PSA = 0.733 / 0.95$

$= 0.772$

$W_{TRIB} \cong 13.1 \times 140' \times 1.25 \; \binom{\text{CONTINUITY}}{\text{FACTOR}}$

$= 2,293 \text{ KIPS}$

$(F_{EQ})_{EL} = 2,293 \times 0.772 = 1,770 \text{ KIPS}$

For preliminary design:

$\qquad F_{EQ} = 1,770 / 1.5 = 1,180 \text{ KIPS}$

Try $1\frac{1}{2}''$ diameter, Grade 55 anchors.

$\qquad A_b = \pi (1.5)^2 / 4 = 1.767 \text{ in}^2 / \text{anchor}$

$\qquad F_{ub} = 75 \text{ KSI} \qquad N_s = 1 \text{ shear plane}$

$\phi P_n = 0.75 \times 0.50 \times 75 \times 1.767 \times 1.0$

$\qquad = 49.7 ^k / \text{anchor}$

$N_{anch} = 1,180 / 49.7 = 23.7 \text{ anchors}$

Use 6 anchors per bay (4 bays)
evenly spaced between girders
$\qquad (1\frac{1}{2}''$ diameter, Grade 55$)$

$$T_r = 2 \times 0.375 + 15 \times 0.75 = 12.00''$$

$$d_L = 8'' \qquad A_{PLUG} = \pi (8)^2 / 4 = 50.3 \, in^2$$

$$F_y = 1.3 \, ksi \times 50.3 \, in^2$$

$$\Rightarrow \underline{F_y = 65.3 \, KIPS}$$

$$A_b = \frac{\pi}{4} (31.5^2 - 8^2) = 729 \, in^2$$

$$k_d = 1.1 \times 0.055 \times 729 / 12.00$$

$$\Rightarrow \underline{k_d = 3.68 \, kips/inch}$$

$$k_i = [21.75 \times 50.3 + 0.055 \times 729] / 12.00$$

$$\Rightarrow \underline{k_i = 94.5 \, kips/inch}$$

$$\Delta_y = 65.3 k / 94.5 \, k/inch$$

$$\Rightarrow \underline{\Delta_y = 0.691 \, inches}$$

$$F_{max} = 65.3 + 3.68 (20.00 - 0.691)$$

$$\Rightarrow \underline{F_{Max} = 136.4 \, KIPS}$$

$$D_r = 0.375 \qquad D_c = 1.00$$

$$Q_d = 65.3 (1 - 3.68/94.5) = 62.8 \, KIPS$$

PROBLEM 8.4	TEH	2/3

$$k_{EFF} = 136.4 / 20.00 = 6.82 \; k/in$$

$$\xi_{EFF} = \frac{2(62.8)(20 - 0.691)}{\pi (20)^2 (6.82)}$$

$$\Rightarrow \xi_{EFF} = 0.283 \; (28.3\% \text{ of critical})$$

$$\gamma_r = \frac{0.375 \; (31.5)^2 (0.010)}{0.75 \times 12.00} = 0.413$$
(all cyclic)

$$\gamma_{s,s} = 3.8/12 = 0.317, \text{ static}$$
$$0.7/12 = 0.058, \text{ cyclic}$$

$$\gamma_{s,eq} = 20/12 = 1.667$$

$$(\gamma_{c,DL} + 0 + 0.317) + 1.75 (\gamma_{c,LL} + 0.413 + 0.058)$$

$$\leq 5.0$$

$$\Rightarrow \gamma_{c,DL} + 1.75 \gamma_{c,LL} \leq 3.859$$

$$\gamma_{c,LL} = \gamma_{c,DL} / 1.68$$

$$\Rightarrow \gamma_{c,DL} \leq 1.890$$

PROBLEM 8.4	TEH	3/3

$$\left[\ \gamma_{c,DL} + \gamma_{c,LL} + 0.5\,(0.413) + (0.317 + 0.058)\right.$$
$$\left. + 1.667\ \right] \leq 5.5$$

$$\gamma_{c,DL} + \gamma_{c,LL} \leq 3.251$$

$$\gamma_{c,DL} + \frac{\gamma_{c,DL}}{1.68} \leq 3.251$$

$$\Rightarrow \gamma_{c,DL} \leq 2.038$$

The controlling limit is:

$$\gamma_{c,DL} \leq 1.890$$

$$S = \frac{31.5^2 - 8^2}{4 \times 31.5 \times 0.75} = 9.82$$

$$\gamma_{c,DL} = \frac{(P_{DL}/729)(1.0)}{0.055\,(9.82)} \leq 1.890$$

$$\Rightarrow P_{DL} \leq 744\ \text{KIPS}$$

$$P_{LL} \leq 744/1.68 = 443\ \text{KIPS}$$

$$\underline{P_{TOTAL} \leq 1,187\ \text{KIPS}}$$

8.6   EXERCISES

E8.1.

For the Project Bridge, estimated reactions for an interior girder are as follows:

$P_{DCI} = 142$ kips per girder

$P_{DC2} = 23$ kips per girder

$P_{DW} = 32$ kips per girder

$P_{LL+IM} = 180$ kips per girder

A fixed bearing (no expansion/contraction movement) is proposed at the pier. A reinforced elastomeric bearing with specified shear modulus $G = 100$ psi is proposed. Bearing dimensions are $L = 12$ inches and $W = 16$ inches. The proposed bearing has three internal elastomer layers, each 0.50 inches thick, and two external elastomer layers, each 0.25 inches thick. Internal reinforcement layers are 0.125-inch-thick A36 steel. The bearing has no holes and bonded steel plate on top only. Live load rotation is 0.002 radians. Dead load rotation is taken to be zero.

Check the strain and stability requirements, check the steel layer thickness, and determine the forces delivered to the pier.

E8.2.

A 30-inch $(W) \times 18$-inch (L) reinforced elastomeric expansion bearing is proposed for a bridge girder with the following loads:

$P_{DC} = 200$ kips

$P_{DW} = 55$ kips

$P_{LL+IM} = 160$ kips

$\theta_{LL} = 0.0035$ radians

$\Delta = 3.50$ inches each way

The bearing is to consist of ½-inch-thick internal elastomer layers with ¼-inch-thick external cover layers and 1/8-inch-thick steel reinforcement layers. Specified elastomer shear modulus is 100 psi. Dead load rotation is accommodated by using beveled sole plates and may be taken to be equal to zero for the design. Determine the minimum number of internal layers based on expansion requirements. Check the bearing for all other criteria.

9 Reinforced Concrete Substructures

Many bridge substructures today are constructed using reinforced concrete. These include bent caps and columns, spread footings, pile caps, drilled shafts, and micropiles. Whereas accelerated bridge construction measures have incorporated precast components, the material presented here for conventionally constructed substructures focuses on cast-in-place components, with the exception of driven concrete piles.

Load and resistance factors for buildings and other structures are not necessarily appropriate for bridge substructures. Load factors for bridges have been discussed in early chapters of this book. A sampling of AASHTO LRFD BDS resistance factors for reinforced concrete elements is summarized in Table 9.1.

Tension-controlled reinforced concrete sections are defined as sections with a net tensile strain in the extreme layer of tensile reinforcement, ε_t, greater than or equal to the tension-controlled strain limit, ε_{tl}, when the concrete strain reaches a value of 0.003.

Compression-controlled reinforced concrete sections are defined as sections with a net tensile strain in the extreme layer of tensile reinforcement, ε_t, less than or equal to the compression-controlled strain limit, ε_{cl}, when the concrete strain reaches a value of 0.003.

The tension-controlled strain limit, ε_{tl}, is determined as follows:

- $\varepsilon_{tl} = 0.005$ for reinforcement with $f_y \leq 75$ ksi
- $\varepsilon_{tl} = 0.008$ for reinforcement with $f_y = 100$ ksi
- ε_{tl} is determined by linear interpolation for reinforcement with $75 < f_y < 100$ ksi

The compression-controlled strain limit, ε_{cl}, is determined as follows:

- $\varepsilon_{cl} = f_y/E_s$, but not > 0.002, for reinforcement with $f_y \leq 60$ ksi
- $\varepsilon_{cl} = 0.004$ for reinforcement with $f_y = 100$ ksi
- ε_{cl} is determined by linear interpolation for reinforcement with $60 < f_y < 100$ ksi

For sections with net tensile strain in the extreme layer of tension reinforcement between ε_{cl} and ε_{tl}, linear interpolation, as given by Equation 9.1, is used to determine the appropriate resistance factor.

$$0.75 \leq \phi = 0.75 + \frac{0.15(\varepsilon_t - \varepsilon_{cl})}{\varepsilon_{tl} - \varepsilon_{cl}} \leq 0.90 \tag{9.1}$$

DOI: 10.1201/9781003265467-9

TABLE 9.1
AASHTO Resistance Factors for Concrete Elements

Condition	Resistance factor, ϕ
Tension-controlled reinforced concrete	0.90
Compression-controlled reinforced concrete	0.75
Shear and torsion	0.90
Bearing on concrete	0.70
Resistance during pile driving	1.00

This chapter provides a discussion of design considerations for pier caps, pier columns, spread footings, pile caps, drilled shafts, and pile bents.

9.1   PIER CAP DESIGN

When the depth of a reinforced concrete flexural member, d_l, is greater than 3 feet, side bars are required. Typical pier caps exceed 3 feet in depth, so side steel, in accordance with Equation 9.2, is often required. See the AASHTO LRFD BDS, Section 5.6.7. This reinforcement is effective in controlling crack widths. The depth, d_l, is defined as the distance from the extreme compression fiber to the centroid of the extreme steel tensile layer. The side bars, having area on each side face equal to A_{sk} (in²/ft), are to be located in each side face for a distance of no less than $d_l/2$ from the tension bars, and at a spacing to exceed neither $d_l/6$ nor 12 inches. The total area of side bars must not exceed 25% of the total tensile reinforcement area.

$$A_{sk} \geq 0.012\left(d_l - 30\right) \tag{9.2}$$

Also, to control cracks, Section 5.6 of the AASHTO LRFD BDS requires that bar spacing in the extreme tension layer of reinforcement must not exceed s as given by Equation 9.3.

$$s \leq \frac{700\gamma_e}{\beta_s f_{ss}} - 2d_c \tag{9.3}$$

$$\beta_s = 1 + \frac{d_c}{0.7\left(h - d_c\right)} \tag{9.4}$$

β_s is the ratio of strain at the extreme tension face to strain in the centroid of the reinforcement layer closest to the tension face. d_c is the distance from the extreme tension fiber to the centroid of the reinforcement closest to the tension face. The overall height of the member is h. The Service limit state stress in the reinforcement is f_{ss}, which is not to exceed 0.60 times f_y. The exposure factor, γ_e, is 1.00 for Class 1 exposure conditions and 0.75 for Class 2 exposure conditions. Class 1 exposure corresponds to an estimated crack width of 0.017 inches. Class 2 exposure corresponds to an estimated crack width of 0.013 inches, as the exposure factor is directly proportional to the estimated crack width. Decks and substructures exposed to water are examples of cases where it may be advisable to use Class 2 exposure for crack control.

The summary of shear provisions in reinforced concrete members presented here is applicable to so-called "B-regions" without significant torsion. These are regions where the assumption that plane sections remain plane is judged to be valid. For deep beams and other disturbed regions ("D-regions") the reader is referred to Section 5.8 of the AASHTO LRFD BDS.

The shear depth, d_v, is taken to be equal to the distance between cross-sectional resultant tensile and compressive forces but must not be less than either $0.72h$ or $0.90d_e$. d_e is the distance from the extreme compression fiber to the resultant cross-sectional tensile force and h is the total member depth. The shear width, b_v, is the minimum web width.

For circular members, d_v may be determined by cross-sectional analysis, or Equation 9.5 from the LRFD BDS Commentary to Section 5.7.2.8 may be used. The column diameter, D, and the diameter of the circle passing through the centroid of the longitudinal reinforcement, D_r, are readily available prior to any design having been completed.

$$d_v = 0.9d_e = 0.9\left(\frac{D}{2} + \frac{D_r}{\pi}\right) \tag{9.5}$$

Nominal shear resistance in non-prestressed members, V_n, consists of contributions from the concrete and from the steel shear reinforcement. Equation 9.6 provides the nominal resistance. For normal-weight concrete with stirrups inclined 90 degrees to the longitudinal axis of the beam, Equations 9.7 and 9.8 provide the concrete and steel contributions, respectively. The units on f'_c must be ksi.

$$V_n = V_c + V_s \leq 0.25 f'_c b_v d_v \tag{9.6}$$

$$V_c = 0.0316\beta\sqrt{f'_c} b_v d_v \tag{9.7}$$

$$V_s = \frac{A_v f_y d_v}{s} \cdot \cot\theta \tag{9.8}$$

The minimum area of shear reinforcement in non-prestressed members, required whenever $V_u > 0.5\phi(V_c)$, is given by Equation 9.9 for normal-weight concrete. The resistance factor for shear is 0.90.

$$A_v \geq 0.0316\sqrt{f'_c} \cdot \frac{b_v s}{f_y} \tag{9.9}$$

Shear reinforcement spacing limits depend on the shear stress on the concrete, v_u, which is calculated using Equation 9.10 for non-prestressed members. The spacing limit is given in Equation 9.11.

$$v_u = \frac{|V_u|}{\phi b_v d_v} \tag{9.10}$$

$$s_{\max} = \begin{cases} 0.4 d_v \le 12 \text{ inches,} & \text{if } v_u \ge 0.125 f_c' \\ 0.8 d_v \le 24 \text{ inches,} & \text{if } v_u < 0.125 f_c' \end{cases} \tag{9.11}$$

Determination of the coefficients β and θ is necessary to accurately assess shear resistance. These factors depend on the net tensile strain in the centroid of the tension reinforcement, ε_s. The discussion here is limited to sections with at least the minimum amount of transverse shear reinforcement. For such cases, the strain and the coefficients are given by Equations 9.12 through 9.14 for non-prestressed members. The steel area, A_s, is the area of longitudinal reinforcement on the flexural tension side of the member only. The strain may also be determined by a cross-sectional analysis. For other cases, the reader is referred to Section 5.7.3.4.2 of the AASHTO LRFD BDS.

$$\varepsilon_s = \frac{\left(\dfrac{|M_u|}{d_v} + 0.5 N_u + |V_u| \right)}{E_s A_s} \tag{9.12}$$

$$\beta = \frac{4.8}{1 + 750 \varepsilon_s} \tag{9.13}$$

$$\theta = 29 + 3500 \varepsilon_s \tag{9.14}$$

For calculated strains, ε_s, less than zero, ε_s may be taken to be equal to zero. The upper limit on ε_s to be used for calculation of shear resistance coefficients is $\varepsilon_s \le 0.006$. This effectively places a lower limit on β equal to 0.873 and an upper limit on θ equal to 50 degrees. Other limitations on the parameters are summarized below.

- N_u, the factored axial force on the cross section, is positive if tension, negative if compression.
- M_u, the factored moment at the section, must not be less than $V_u \times d_v$.
- For sections closer than d_v to the face of support, ε_s at a distance d_v from the face of support may be used to determine β and θ, unless a concentrated load is located within d_v from the support, in which case ε_s should be determined at the face of support.

For members with small M_u/V_u ratios, relative to member depth, it may be advisable to use detailed, complex section analysis, including flexure-shear interaction. Problem 9.5 demonstrates the significant difference such an analysis may make in estimating design resistance.

For additional shear-related longitudinal reinforcement requirements, refer to Section 5.7.3.5 of the AASHTO LRFD BDS.

9.2   PIER COLUMN DESIGN

Column reinforcement must typically be in the range of 1% to 8% of the gross column area. Equation 9.15 provides the minimum steel area, irrespective of any seismic design considerations, from the LRFD BDS, Section 5.6.4.2.

$$A_s \geq 0.135 A_g \left(\frac{f'_c}{f_y} \right) \tag{9.15}$$

However, for Seismic Zone 2, column longitudinal reinforcement must be between 1% and 6% of the gross cross-sectional area. For Seismic Zones 3 and 4, the longitudinal column reinforcement must be between 1% and 4% of the gross concrete area.

For columns not considered braced against sidesway, slenderness effects may be ignored whenever Kl_u/r is less than 22. For columns considered braced against sidesway, slenderness may be ignored whenever Kl_u/r is less than 34-12 (M_1/M_2). M_1 and M_2 are the smaller and larger end moments, respectively, and the ratio of the two is positive when the column is in single curvature, negative if the column is in double curvature. Approximate second-order analysis is acceptable for the evaluation of slenderness effects whenever Kl_u/r is less than 100. Equation 9.16 is an acceptable estimate for the flexural rigidity, EI. The ratio of factored permanent moment to factored total moment, β_d, appears in the equation and requires a measure of engineering judgment. Equation 9.17 provides the critical buckling load. Equations 9.18 and 9.19 give the required amplifiers on column moments producing no appreciable sidesway (δ_b) and on column moments resulting in appreciable sidesway (δ_s).

$$EI = \frac{E_c I_g}{2.5 \left(1 + \beta_d \right)} \tag{9.16}$$

$$P_e = \frac{\pi^2 EI}{\left(Kl_u \right)^2} \tag{9.17}$$

$$\delta_b = \frac{1}{1 - \dfrac{P_u}{0.75 P_e}} \tag{9.18}$$

$$\delta_s = \frac{1}{1 - \dfrac{\Sigma P_u}{0.75 \Sigma P_e}} \tag{9.19}$$

$$M_u = \delta_b M_{ub} + \delta_s M_{us} \tag{9.20}$$

Factored axial resistance of non-prestressed concrete columns, ϕP_n, may be determined from Equation 9.21 for columns with spiral reinforcement, or Equation 9.22 for tied columns. The factor, k_c, is equal to 0.85 for concrete strengths up to and including 10 ksi, 0.75 for concrete strengths 15 ksi or more, and linear interpolation is used for intermediate concrete strength values.

$$\phi P_n = \phi \times 0.85 \left[k_c f_c' \left(A_g - A_{st} \right) + f_y A_{st} \right] \tag{9.21}$$

$$\phi P_n = \phi \times 0.80 \left[k_c f_c' \left(A_g - A_{st} \right) + f_y A_{st} \right] \tag{9.22}$$

Biaxial flexure due to eccentricity of axial load effects may be determined from single-axis eccentricities using the following formulas. Equation 9.23 is applicable whenever the factored axial load, P_u, is less than $0.10 f_c' A_g$. Equation 9.24 applies otherwise.

$$\frac{M_{ux}}{\phi M_{nx}} + \frac{M_{uy}}{\phi M_{ny}} \leq 1.00 \tag{9.23}$$

$$\frac{1}{\phi P_{nxy}} = \frac{1}{\phi P_{nx}} + \frac{1}{\phi P_{ny}} + \frac{1}{\phi P_o} \tag{9.24}$$

$$\phi P_o = \phi \left[k_c f_c' \left(A_g - A_{st} \right) + f_y A_{st} \right] \tag{9.25}$$

For circular columns of diameter D, and with a reinforcing ring of diameter D_r, the effective shear depth, d_v, may be determined from Equation 9.26.

$$d_v = 0.90 \left(\frac{D}{2} + \frac{D_r}{\pi} \right) \tag{9.26}$$

Column design generally requires criterion checks at both the minimum and maximum axial load levels, and at multiple Strength limit states.

9.3   SPREAD FOOTING DESIGN

Footing design provisions in the AASHTO LRFD BDS are found in Chapter 5 (Section 5.12.8) and Chapter 10 (Section 10.6).

For square footings, the LRFD BDS in Section 5.12.8 requires footing reinforcement to be uniformly distributed across the footing width in each direction. For rectangular footings, long bars are to be distributed uniformly across the short dimension, and short bars are to be distributed along the long dimension as follows.

A central band, equal in width to the short dimension of the footing, is to have a uniformly distributed area of steel equal to $A_{s\text{-}BW}$. The remainder of the required steel is to be uniformly distributed on each side of the central band. The total steel area required by analysis for the short bars is denoted $A_{s\text{-}SD}$. Equation 9.27 gives the area of steel required within the central band. The parameter, β, is equal to the ratio of the long dimension to the short dimension.

$$A_{s-BW} = A_{s-SD} \left(\frac{2}{\beta + 1} \right) \tag{9.27}$$

The critical section for moment in footings is to be taken at the face of the column. The critical section for one-way shear is taken at d_v from the face of the column. The critical section for two-way shear is taken as a perimeter at $d_v/2$ from the column or concentrated load. An exception occurs for any loading conditions which place the top of the footing in tension at the column interface. The critical shear for such a condition is at the face of the column.

One-way shear resistance for footings may often be computed using Equation 9.7 with the shear coefficient $\beta = 2$. Refer to the AASHTO LRFD BDS Section 5.7.3.4.1 for cases where this may not be acceptable.

Two-way shear resistance for footings constructed with normal-weight concrete is given in the AASHTO LRFD BDS in Section 5.12.8.6.3 and here in Equation 9.28. The perimeter, b_o, is taken to be $d_v/2$ from the concentrated load. The ratio of the long side to the short side of the concentrated load area gives the parameter β_c.

$$V_n = \left(0.063 + \frac{0.126}{\beta_c}\right)\sqrt{f'_c}b_o d_v \leq 0.126\sqrt{f'_c}b_o d_v \qquad (9.28)$$

If V_u is greater than or equal to ϕV_n, then either (a) the footing thickness must be increased, (b) the concrete strength must be increased, or (c) shear reinforcement must be added. Shear resistance from shear reinforcement in footings is to be computed by Equation 9.8 with $\theta = 45$ degrees.

Section 10.6.5 of the LRFD BDS requires that a trapezoidal stress distribution beneath footings be used for structural design.

Eccentricity limits for footings with axial load combined with moment are specified for both the Strength and Extreme Event limit states and are as follows:

- Strength limit state eccentricity $\leq 1/3$ of the footing dimension (soil)
- Strength limit state eccentricity ≤ 0.45 times the footing dimension (rock)
- Extreme Event limit state eccentricity $\leq 1/3$ of the footing dimension for $\gamma_{EQ} = 0.00$ (load factor on live load at the Extreme Event limit state)
- Extreme Event limit state eccentricity ≤ 0.40 times the footing dimension for $\gamma_{EQ} = 1.00$ (load factor on live load at the Extreme Event limit state)

For seismic design in accordance with the AASHTO LRFD GS, footing dimensions must satisfy $(L - D_c)/(2H_f) \leq 2.5$. Otherwise, the footing may not be classified as rigid and special analyses are required. The footing length in the direction of loading is L, the column diameter of depth in the detection of loading is D_c, and the footing depth is H_f. Joint shear in footings must also be assessed for bridges in Seismic Design Categories C and D in accordance with Section 6.4.5 of the LRFD GS.

9.4 PILE CAP DESIGN

Pile cap design requires determination of cap reinforcement and determination of the maximum and minimum pile loads at the Strength and Service limit states. Figure 9.1 depicts an example of pile cap and pile configuration. The vertical axial

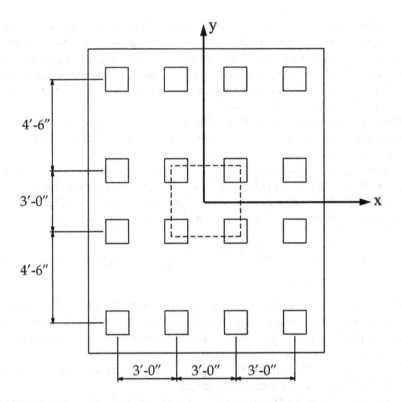

FIGURE 9.1   Pile cap example.

load in any pile, p_i, may be calculated using Equation 9.29. P_u is the total vertical load (including self-weight of the pile cap with appropriate load factor) applied at the center of the pile cap (indicated by the dotted line, representing a column, in the figure). M_{ux} is the moment about the x-axis and M_{uy} is the moment about the y-axis. The maximum pile load occurs at the corner for which the second and third terms in Equation 9.29 take on the positive sign. The minimum pile load occurs at the corner for which the second and third terms take on the negative sign. With compression indicated by a positive pile load, negative results indicate tension (uplift) on the pile in question.

$$p_i = \frac{P_u}{N} \pm \frac{M_{ux} \cdot y_i}{\sum\limits_{i=1}^{N} y_i^2} \pm \frac{M_{uy} \cdot x_i}{\sum\limits_{i=1}^{N} x_i^2} \tag{9.29}$$

Although some portion of the applied vertical load is likely carried by the soil underneath the pile cap, the LRFD BDS, in Section 5.12.9, requires that all loads resisted by the pile cap and its self-weight are to be assumed to be transmitted to the piles.

Piling may be steel H-piles, precast concrete piles, or steel pipe piles.

Section 4 of the AASHTO LRFD BCS contains numerous requirements which need to be considered in addition to design requirements in the AASHTO LRFD BDS.

Open-end pipe piles may be filled with concrete. ASTM A252 Grade 2 is used for pipe piles. Although ASTM A252 is common, structures with seismic requirements may require additional qualification. One example of this from the AASHTO LRFD BCS is as follows:

> "Pipe shall be ASTM A252, but dimensional tolerance as per API 5L and elongation of 25 percent minimum in 2.0 in. The carbon equivalency shall not exceed 0.05 percent".

API 5L could be specified, but it requires hydrostatic testing and 48.0 in. outside diameter is the largest diameter covered by API 5L.

The AASHTO LRFD BCS requires that driving stresses for steel piles be limited to 90% of yield. One rule-of-thumb means for satisfying this is to limit required Service limit state pile loads to 25% of yield, though, strictly speaking, driving stresses are to be determined by wave equation analysis. Driving stresses for concrete piles are limited to $0.85f'_c$ minus the effective prestress, if prestressed piles are used.

Predrilling of a hole with a continuous flight auger or a wet rotary bit may be required to install piling. Predrilling may be beneficial when driving the pile will displace the upper soil enough to push adjoining piles out of the proper position, or to limit vibration in the upper layers. Predrilled holes are typically smaller than the diameter or diagonal of the pile cross-section and sufficient to allow penetration of the pile to the required elevation. If subsurface obstructions are encountered, then the diameter of the predrilled hole may be increased to facilitate pile installation or to avoid obstructions. Jetting is another method which can be used to facilitate driving.

Ends of closed-end pipe piles must be closed with a flat plate or a forged or cast steel conical point, or other end-closure of approved design. End plates must be at least ¾ inches thick, cut flush with the outer pile wall. The end of the pipe must be beveled prior to welding to the end plate with a partial penetration groove weld.

Pipe piles are sometimes filled with concrete, depending on the design strategy employed. Before concrete is placed in a pipe pile, the pile must be inspected to confirm the full pile length and a dry bottom condition. Any water accumulations in the pipe piles is to be removed before the concrete is placed. The minimum compressive strength of concrete used for concrete-filled pipe piles is 2.5 ksi or that required by design, whichever is larger. A slump of not less than 6 inches and not more than 10 inches must be used for the concrete. Concrete must be placed in each pile in a continuous operation. No concrete is to be placed in a pile until all driving within a radius of 15 feet of the pile has been completed, or all driving within the 15-foot radius must be discontinued until the concrete in the last pile cast has set for at least two days.

In accordance with Section 5.12.9 of the AASHTO LRFD BDS, piles are to be embedded at least 12 inches into the pile cap, as specified in Section 10.7.1.2. For concrete piles, anchorage reinforcement must be provided by either (a) an extension of the pile reinforcement or (b) the use of dowels. Uplift forces or stresses induced by flexure shall be resisted by the reinforcement. The steel ratio for the anchorage reinforcement must be no less than 0.005, and the number of bars must be at least four. While Section 5.12.9 requires that the anchorage reinforcement be developed sufficiently to resist a force of 1.25 $f_y A_s$, arguments have been made for omitting anchorage reinforcement, even when uplift is indicated, and eliminating the tension pile from subsequent load distribution calculations for the load case considered (i.e., re-compute the centroid of the pile group, ignoring the tension piles). This does seem to be a reasonable design approach.

Pile driving equations available for estimating pile resistance from driving logs include the Gates method and the Engineering News Record (ENR) method. Resistance factors are $\phi = 0.40$ for the Gates method and $\phi = 0.10$ for the ENR method. Equation 9.30 is the Gates formula. Equation 9.31 is the ENR formula. The resulting nominal load, R_{ndr}, is in kips.

$$R_{ndr} = 1.75\sqrt{E_d} \, \log_{10}\left(10N_b\right) - 100 \qquad (9.30)$$

$$R_{ndr} = \frac{12E_d}{s + 0.1} \qquad (9.31)$$

Although both the Gates and ENR methods are presented in the LRFD BDS, the units vary between the methods.

- For the Gates formula, E_d must be expressed in units of ft-lbs.
- For the ENR formula, E_d must be expressed in units of ft-kips.
- In the Gates formula, N_b is the number of blows per inch of pile set. Recall that geotechnical testing typically reports blow counts in blows per foot.
- In the ENR formula, $s = 1/N_b$, is the pile set in inches per blow.

For additional pile requirements, see Sections 9.6 and 9.9 of this text.

9.5   DRILLED SHAFT DESIGN

A partial set of AASHTO LRFD BDS requirements for geotechnical design of drilled shafts is summarized below in Equations 9.32 and 9.33 for nominal tip and side resistance, respectively. With intact rock at least two shaft diameters below the shaft tip under normal conditions, full tip resistance (q_p) and side resistance (q_s) for shafts socketed into rock are given by the following equations.

$$q_p = 2.5q_u \qquad (9.32)$$

$$q_s = p_a\sqrt{\frac{q_u}{p_a}} \qquad (9.33)$$

q_u is the uniaxial compressive strength of the rock in ksf.

p_a is atmospheric pressure, taken as 2.12 ksf.

Resistance factors at the Strength limit state for static analysis of drilled shaft design geotechnical resistance vary from 0.40 for tip resistance in clay up to as high as 0.55 for side resistance in sand or rock. The reader is referred to Chapter 10 of the AASHTO LRFD BDS for a complete discussion of applicable resistance factors.

The geotechnical engineer may provide permissible percentages of tip and side resistance in combination to determine total geotechnical resistance. For example, the engineer may determine that 50% of the tip resistance may be directly combined with 100% of the side resistance (or *vice versa*, or any other combination of percentages) to establish the total design geotechnical resistance.

The top portion of a drilled shaft socket is sometimes ignored in side-resistance computations, again as directed by the geotechnical engineer.

9.6 PILE BENT DESIGN

Pile bents are often an economical option for bridges. Pile bents make use of piles extended to the pier cap and embedded therein, rather than completely in-ground piles. Figure 9.2 depicts the elements of a typical pile bent. Span lengths are limited

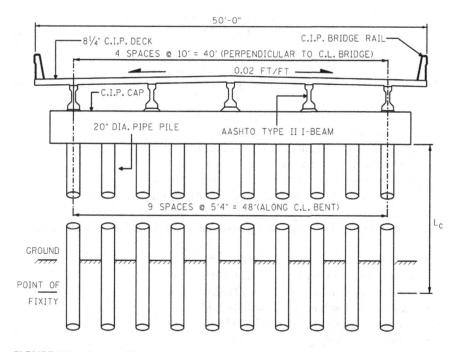

FIGURE 9.2 Typical pile bent.

by the ability of local constructors to provide, drive, and load-test the piling. The larger the pile type and driving capacity, the longer will be the span for which a pile bent bridge proves economical.

For Strength and Service limit states, pile bent design requires flexure-axial load interaction criteria. Unlike completely in-ground piles, which are often designed as purely axial elements, the piles in a pile bent are subjected to significant flexure.

The piles in a pile bent may be precast, reinforced concrete, precast, prestressed concrete, steel HP sections, or steel pipe piles. When pile lengths longer than about 75 ft are expected, it may be advisable to specify pipe piles and to provide splicing details to develop the full resistance of the pile. Certainly, it is necessary to ensure that splice locations are far away from anticipated plastic hinge locations for seismic design.

While hollow, un-filled pipe piles have been used successfully, it may be preferable to use concrete-filled steel tubes when pipe piles in regions of high seismic hazard are encountered.

Section 9.9 in this text presents a detailed treatment of pipe pile and concrete-filled steel tube (CFST) design requirements.

An estimate of the depth to point of fixity is required for the analysis and design of pile bents. Section 10.7.3.13.4 of the AASHTO LRFD BDS presents, in the commentary, estimates of depth to fixity, in feet, for both clays (Equation 9.34) and sands (Equation 9.35).

$$D_f = 1.4 \left[\frac{E_p I_w}{E_s} \right]^{0.25} \tag{9.34}$$

$$D_f = 1.8 \left[\frac{E_p I_w}{n_h} \right]^{0.20} \tag{9.35}$$

- E_p = modulus of elasticity of the pile, ksi
- I_w = weak axis moment of inertia for pile, ft^4
- E_s = soil modulus for clays = 0.465 S_u, ksi
- S_u = undrained shear strength of clays, ksf
- n_h = rate of increase of soil modulus with depth, ksi/ft

Representative ranges of values for n_h, based on broad classification, are as follows:

- loose sand, dry or moist, n_h = 0.417 ksi/ft
- loose sand, submerged, n_h = 0.208 ksi/ft
- medium sand, dry or moist, n_h = 1.110 ksi/ft
- medium sand, submerged, n_h = 0.556 ksi/ft
- dense sand, dry or moist, n_h = 2.780 ksi/ft
- dense sand, submerged, n_h = 1.390 ksi/ft

For clays, E_s ranges, again based on broad categorization, are as follows:

- medium stiff clay, $E_s = 0.347$ to 2.08 ksi
- stiff clay, $E_s = 2.08$ to 6.94 ksi
- very stiff clay, $E_s = 6.94$ to 13.89 ksi

Undrained shear strength, S_u, generally falls within the following ranges:

- soft soil, $S_u < 1$ ksf
- stiff soil, $S_u = 1$–2 ksf
- very stiff soil, S_u exceeds 2 ksf

Other rough estimates of depth to fixity are given in the literature (Priestley et al., 1996). For moment, the depth to fixity has been estimated at approximately 1 to 2 diameters. For deflection, the estimate is 4 to 5 diameters.

Section 5.12.9.3 of the LRFD BDS requires that concrete piles, whether reinforced or prestressed, must have a cross-sectional area no less than 140 in² unless exposure to saltwater is possible, in which case the minimum cross-sectional area is 220 in². Concrete used for prestressed piles must have a specified 28-day concrete strength no less than 5 ksi.

For prestressed concrete piling, the uniform compressive stress on the cross section, after prestress losses, must be no less than 0.700 ksi.

For bridges in LRFD BDS Seismic Zone 1 (LRFD GS Seismic Design Category A), transverse, spiral reinforcement is required over the entire pile length for prestressed concrete piles, whether round, rectangular, or other. The main spiral reinforcement must be spiral wire not less than W3.9 at a pitch of no more than 6 inches for piles having the least dimension no greater than 24 inches. For larger piles, the required transverse spiral is to be no less than W4.0 wire with a pitch not to exceed 4 inches. See Section 5.12.9.4.3 of the LRFD-BDS for spacing limits near the ends of piles.

For LRFD BDS Seismic Zones 2, 3, and 4 (LRFD GS Seismic Design Categories B, C, and D), enhanced requirements must be satisfied for piles. Transverse reinforcement must be at least No. 3 bar at a pitch not to exceed 9 inches.

For Seismic Zones 3 and 4, the upper end of the pile must contain confinement reinforcement as a potential plastic hinge location. The potential hinge region extends from the bottom of the pile cap for a distance no less than either (a) two pile diameters or (b) 24 inches.

For pile bents, the top potential hinge region extends from the bottom of the cap for a distance no less than the greater of (a) the maximum cross-section dimension, (b) one-sixth of the clear height, or (c) 18 inches. For in-ground hinges in pile bents, the bottom potential hinge region extends from three pile diameters below the point of maximum, in-ground moment to one diameter above the maximum moment point. However, the top limit for the in-ground potential hinge region shall not be less than 18 inches above the top of the ground.

For circular columns and piles, Equation 9.36 provides the required volumetric spiral or seismic hoop reinforcement for plastic hinge regions. The core diameter, d_c, is measured to the outside diameter of the spiral or hoop. The pitch, s, is measured vertically to the center of the spiral or hoops. A_{sp} is the cross-sectional area of the spiral or hoop bar. The specified minimum yield strength, f_{yh}, is that for the spiral or hoop bar, not necessarily the same as that for longitudinal bars. The specified concrete strength, f'_c, is used for the calculations, rather than the expected concrete strength, f'_{ce}. Certainly, Equation 9.37 from Section 5.6.4.6 of the LRFD BDS, for non-seismic confinement requirements, should always be checked as well. The core area, A_c, is based on the diameter to the outside of the spiral or hoop as well.

Equations 9.38 and 9.39 provide the required confinement reinforcement, A_{sh}, within a spacing equal to s, for rectangular piles and columns. The dimension, h_c, is measured to the outside of the transverse bars in the direction of loading. The spacing, s, must exceed neither (a) 4 inches, nor (b) one-quarter of the least cross-sectional dimension.

$$\rho_s = \frac{4A_{sp}}{d_c s} \geq 0.12 \frac{f'_c}{f_{yh}} \tag{9.36}$$

$$\rho_s = \frac{4A_{sp}}{d_c s} \geq 0.45 \left(\frac{A_g}{A_c} - 1 \right) \frac{f'_c}{f_{yh}} \tag{9.37}$$

$$A_{sh} \geq 0.12 s h_c \frac{f'_c}{f_{yh}} \tag{9.38}$$

$$A_{sh} \geq 0.30 s h_c \left(\frac{A_g}{A_c} - 1 \right) \frac{f'_c}{f_{yh}} \tag{9.39}$$

For either round or rectangular columns and piles, the yield strength of the transverse reinforcement, f_{yh}, is the minimum specified value for the reinforcement used, but must not exceed 75 ksi.

For pile bent design by the displacement-based provisions of the LRFD GS, the displacement ductility demand, μ_D, is not to exceed 4. For single column bents, μ_D is limited to 5, and for multi-columns bents with above-ground hinging, μ_D is limited to 6.

For slenderness effects, it is often necessary to determine an appropriate effective length factor, K, from section 5.6.2.5 of the LRFD BDS. When integral abutments are used at both ends of the bridge (i.e., no expansion joints), it may be possible to treat pile bents as braced out-of-plane ($K = 1.0$) and as rigid frames in-plane ($K = 1.2$). Provided sufficient stiffness is available in the longitudinal direction from the integral abutments, the assumption may or may not be valid. With expansion abutments, it would most likely seem advisable to treat the pile bent as unbraced out-of-plane ($K = 2.0$).

9.7 BRIDGE PIER DISPLACEMENT CAPACITY UNDER SEISMIC LOADING

Bridge design for seismic effects has increasingly become displacement-based, as opposed to the original, decades-old, force-based provisions. Pushover analyses by computer modeling and approximate hand calculations are both useful.

Approximate pushover analysis of concrete piers calculated by hand may be accomplished using an analysis incorporating the Mander model for confined concrete along with procedures outlined in the literature (Priestley et al., 2007) and reproduced here in Equations 9.40 through 9.48. Displacement estimates may be done by computer modeling or by approximate hand-calculation-based equations, such as those presented here in Equations 9.49 through 9.56.

$$\phi_y \cong \begin{cases} 2.25\left(\varepsilon_y\!\big/_D\right), & \text{circular columns} \\[2ex] 2.10\left(\varepsilon_y\!\big/_D\right), & \text{rectangular columns} \end{cases} \tag{9.40}$$

$$\frac{c}{D} \cong 0.20 + 0.65\frac{P}{f_{ce}'A_g} \tag{9.41}$$

$$\varepsilon_{cu} = \begin{cases} 0.004 + \dfrac{1.4\rho_v f_{yh}\varepsilon_{su}}{f_{cc}'}, & \text{damage control limit state} \\[2ex] 0.004, & \text{serviceability limit state} \end{cases} \tag{9.42}$$

$$\varepsilon_{su} = \begin{cases} 0.06 \text{ to } 0.09, & \text{damage control limit state} \\ 0.015, & \text{serviceability limit state} \end{cases} \tag{9.43}$$

$$\rho_{cc} = \frac{A_s}{A_c} = \frac{\text{Area of longitudinal reinforcement}}{\text{Area of core enclosed by centerlines of hoop or spiral}} \tag{9.44}$$

$$\rho_v = \begin{cases} \dfrac{A_{sp}\pi d_s}{\dfrac{\pi}{4}d_s^2 s} = \dfrac{4A_{sp}}{d_s s} \rightarrow \text{circular hoops or spirals} \\[3ex] \rho_{vx} + \rho_{vy} = \dfrac{A_{sx}}{sh_{cy}} + \dfrac{A_{sy}}{sh_{cx}} \rightarrow \text{rectangular hoops} \end{cases} \tag{9.45}$$

$$f_{cc}' = f_{co}'\left(-1.254 + 2.254\sqrt{1+\frac{7.94f_l'}{f_{co}'}} - 2\frac{f_l'}{f_{co}'}\right) \tag{9.46}$$

$$k_e = \begin{cases} \dfrac{\left(1-\dfrac{s'}{2d_s}\right)}{1-\rho_{cc}} \rightarrow \text{circular hoop effectiveness coefficient} \\[3ex] \dfrac{1-\dfrac{s'}{2d_s}}{1-\rho_{cc}} \rightarrow \text{circular spiral effectiveness coefficient} \\[3ex] \dfrac{\left(1-\sum\dfrac{(w_i')^2}{6b_cd_c}\right)\left(1-\dfrac{s'}{2b_c}\right)\left(1-\dfrac{s'}{2d_c}\right)}{1-\rho_{cc}} \rightarrow \text{rectangular hoop coefficient} \end{cases} \quad (9.47)$$

$$f_l' = \frac{1}{2}k_e\rho_v f_{yh} \rightarrow \text{lateral confining stress on concrete} \quad (9.48)$$

$$\Delta_y = \frac{1}{3}\phi_y\left(L_c + L_{SP}\right)^2 \quad (9.49)$$

$$\Delta_P = \left(\phi_u - \phi_y\right)L_P\left(L_c + L_{SP} - \frac{L_P}{2}\right) \quad (9.50)$$

$$k = 0.20\left(\frac{f_u}{f_y} - 1\right) \leq 0.08 \quad (9.51)$$

$$L_P = kL_c + L_{SP} \geq 2L_{SP} \quad (9.52)$$

$$L_{SP} = 0.15f_{ye}d_{bl} \quad (9.53)$$

$$\Delta_u = \Delta_y + \Delta_P \quad (9.54)$$

$$\mu = \frac{\Delta_u}{\Delta_y} \quad (9.55)$$

$$\phi_u = \text{Min}\begin{cases} \dfrac{\varepsilon_{cu}}{c} \\[2ex] \dfrac{\varepsilon_{su}}{d-c} \end{cases} \quad (9.56)$$

- L_c is the distance from the critical section to the point of contra-flexure
- L_{SP} is the strain penetration distance
- L_P is the plastic hinge length
- d is the column depth
- c is the distance from the compression face of the column to the neutral axis

- d_s is the hoop or spiral diameter measured to the center of the hoop or spiral
- s is the hoop or spiral pitch measure to the center of the spiral or hoop
- s' is the clear hoop or spiral pitch
- A_{sp} is the area of the spiral bar
- f'_{co} is the specified concrete strength at 28 days
- b_c is the out-to-out width measured to the center of the rectangular hoop
- d_c is the out-to-out height measured to the center of the rectangular hoop
- w' is the clear spacing between adjacent longitudinal bars
- ε_{cu} is the ultimate concrete compressive strain for a particular limit state
- ε_{su} is the ultimate steel tensile strain for a particular limit state
- ϕ_y is the section yield curvature
- ϕ_u is the section ultimate curvature
- Δ_y is the lateral yield displacement occurring over the length, L_C
- Δ_P is the lateral plastic displacement occurring over the length, L_C

For rigid frame behavior with plastic hinging at the top and bottom of the column, L_C is one-half of the column height and displacements (Δ_y and Δ_P) are twice that given by the equations above. This does not apply to the implicit displacement equations given below, as the rigid frame versus cantilever behavior is incorporated into the factor, Λ, in the implicit equations.

For Seismic Design Categories B and C, approximate displacement capacity equations are provided in the AASHTO LRFD GS (AASHTO, 2011). The expressions are presented here in Equations 9.57, 9.58, and 9.59. For Seismic Design Category B, Equation 9.57 is applicable. For Seismic Design Category C, Equation 9.58 is applicable.

$$\Delta_C^L = 0.12 H_o \left(-1.27 \ln(x) - 0.32\right) \geq 0.12 H_o \qquad (9.57)$$

$$\Delta_C^L = 0.12 H_o \left(-2.32 \ln(x) - 1.22\right) \geq 0.12 H_o \qquad (9.58)$$

$$x = \frac{\Lambda B_o}{H_o} \qquad (9.59)$$

- $\Lambda = 1$ for fixed–free column end conditions (cantilever)
- $\Lambda = 2$ for fixed–fixed column end conditions (rigid frame)
- B_o is the column diameter for circular columns, or the dimension parallel to the direction in which displacement capacity is being calculated, for rectangular columns, feet
- H_o is the clear column height, feet

Note that H_o and B_o are in feet, whereas the displacement capacity from the equations is in inches.

The displacement capacity equations are intended for use with reinforced concrete column substructures with a minimum clear height of 15 feet. Attempts to use the equations for other situations will likely result in serious error.

9.8   THE ALASKA PILE BENT DESIGN STRATEGY

One innovative design concept for pile bent bridges has been developed by the Alaska Department of Transportation and Public Facilities based on research conducted at the University of California at San Diego (Silva and Seible, 1999).

The method relies on a reinforced concrete section at the cap soffit, with the steel tube providing only external confinement to the reinforced concrete. For the in-ground hinge, the cross section consists of the filled tube. Proof of the method has been provided in research (Silva and Sritharan, 2011). The strategy is particularly effective in minimizing damage to the cast-in-place cap. A two-inch gap is required between the top of the steel tube and the cap soffit. The Washington State Department of Transportation (WSDOT, 2020) has also recognized this alternative design strategy.

Should such a strategy be adopted for a particular project, then the literature should be referenced, particularly with regard to required joint reinforcement.

9.9   CONCRETE FILLED STEEL TUBES (CFST)

The design of concrete-filled steel tubes (CFST) is covered in both the AASHTO LRFD BDS and the AASHTO LRFD GS.

Section 6.6.6.1 of the AASHTO LRFD BDS requires that CFST members expected to experience plastic hinging should be designed according to the AASHTO LRFD GS. However, Section 7.6 of the AASHTO LRFD GS requires that CFST expected to experience plastic hinging be designed according to the AASHTO LRFD BDS, Sections 6.9.2.2, 6.9.5, and 6.12.3.2.2, in addition to the GS requirements.

CFST are used for piles, drilled shafts, columns, and other members subject to axial compression or axial compression and flexure. In the AASHTO LRFD BDS, the design of composite CFST may be performed in accordance with either:

- Section 6.9.6 along with Section 6.12.2.3.3 for composite CFST design, or
- Section 6.9.5 along with Section 6.12.2.3.2 for partially composite CFST design

The provisions specified in LRFD-BDS Section 6.9.6 incorporate a great wealth of research performed since the development of Section 6.9.5 and tend to reduce uncertainty and increase the accuracy of analytical results. Nonetheless, it appears that AASHTO recommends GS Section 7.6 in combination with BDS Sections 6.9.2.2, 6.9.5, and 6.12.3.2.2 for CFST design with plastic hinging.

For Seismic Design Categories C and D, ductile concrete-filled steel pipe, as defined in Section 7.6 shall be made of steels satisfying the requirements of either (a) ASTM A 53 Grade B or (b) API 5L X52. For ASTM A 53 Grade B tubes, the expected yield stress, F_{ye}, is to be 1.5 times the nominal yield stress for overstrength calculations. For API 5L X52 tubes, the expected yield stress is 1.2 times the nominal yield stress.

Overstrength moments are to be determined using expected yield stress for the tube, expected concrete strength ($1.3f'_c$), and an overstrength factor, λ_{mo}, applied to the calculated moment. The overstrength factor is 1.2 for structural steel, 1.2 for reinforced concrete using A706 reinforcing bars, and 1.4 for reinforced concrete using A615 reinforcing bars.

9.9.1 CFST DESIGN IN ACCORDANCE WITH BDS SECTIONS 6.9.6 AND 6.12.2.3.3

Requirements on tube dimensions and concrete fill given in Equations 9.60 and 9.61 are applicable when Section 6.9.6 of AASHTO LRFD BDS is adopted for CFST design. The outside tube diameter is D, the tube thickness is t, E is Young's modulus (for steel tubes, $E = 29,000$ ksi), F_{yst} is the specified minimum yield strength of the tube material.

$$\frac{D}{t} \leq 0.15\frac{E}{F_{yst}} \tag{9.60}$$

$$f_c' \geq \text{Max}\left\{3.0 \text{ ksi}, 0.075F_{yst}\right\} \tag{9.61}$$

The resistance factor, ϕ_c, for combined axial compression and flexure, is 0.90 for CFST design. A nominal P–M interaction diagram, reduced by ϕ_c is used to assess the adequacy of the CFST subjected to applied Strength limit state loads, P_u and M_u. Development of the nominal P–M–interaction curve requires initial determination of axial resistance, ϕP_n. The axial resistance is determined using Equations 9.62 through 9.66. The area of concrete fill is A_c. The cross-sectional area of the tube is A_{st}, with corresponding yield strength F_{yst}. The area of internal reinforcing bars is A_{sb}, with corresponding yield strength, F_{yb}. The un-factored axial load is P.

$$P_o = 0.95f_c'A_c + F_{yst}A_{st} + F_{yb}A_{sb} \tag{9.62}$$

$$EI_{eff} = EI_{st} + EI_{si} + C'E_cI_c \tag{9.63}$$

$$C' = 0.15 + \frac{P}{P_o} + \frac{A_{st} + A_{sb}}{A_{st} + A_{sb} + A_c} \leq 0.90 \tag{9.64}$$

$$P_e = \frac{\pi^2 EI_{eff}}{(KL)^2} \tag{9.65}$$

$$P_n = \begin{cases} \left[0.658^{(P_o/P_e)}\right]P_o, & \text{if } \dfrac{P_e}{P_o} > 0.44 \\[2mm] 0.877P_e, & \text{if } \dfrac{P_e}{P_o} \leq 0.44 \end{cases} \tag{9.66}$$

Material-based P–M interaction may be established using either (a) the plastic stress distribution method (PSDM) or (b) the strain compatibility method (SCM). The

PSDM is detailed in AASHTO LRFD BDS, Section 6.12.2.3.3. The interested reader is directed to this Section of the AASHTO LRFD BDS for further guidance.

Shear resistance of the concrete-filled steel tubes is taken to be equal to the shear resistance of the tube alone, according to the AASHTO LRFD BDS, Section 6.12.3.2.2. See Section 9.9.3 of this text as well.

9.9.2   CFST DESIGN BY BDS SECTIONS 6.9.5 AND 6.12.3.2.2 AND GS SECTION 7.6

Given the difficulty in interpreting BDS Section 6.9.6, combined with the recommendation in the AASHTO LRFD GS that BDS Section 6.9.5, in combination with GS Section 7.6, be used for CFST design, a more detailed discussion of this method is presented here.

The wall thickness of CFST designed using BDS 6.9.5 must satisfy Equation 9.67. The concrete must satisfy Equation 9.68. The specified yield strength of the tube and any reinforcement is not to exceed 60 ksi. The tube area should equal or exceed 4 percent of the total cross-sectional area, or the member should be designed as a reinforced concrete column using Section 5 of the LRFD BDS.

$$\frac{D}{t} \le 0.11 \frac{E}{F_{yst}} \tag{9.67}$$

$$3.0 \text{ ksi} \le f_c' \le 8.0 \text{ ksi} \tag{9.68}$$

Compressive resistance in Section 6.9.5 is taken to be equal to that given by Equations 9.69 through 9.72. When moment magnification is used to estimate second-order effects, Equation 9.73 is applicable for the computation of the Euler buckling load. Flexural resistance is given by Equation 9.74. The plastic moment, M_{ps}, is for the steel tube alone. The yield moment, M_{yc}, is for the composite section. The modular ratio, $n, = E_s/E_c$. The radius of gyration, r_s, in Equation 9.71 is that of the tube alone.

$$F_e = F_y + F_{yr}\left(\frac{A_r}{A_s}\right) + 0.85 f_c'\left(\frac{A_c}{A_s}\right) \tag{9.69}$$

$$E_e = E\left[1+\left(\frac{0.40}{n}\right)\left(\frac{A_c}{A_s}\right)\right] \tag{9.70}$$

$$\lambda = \left(\frac{KL}{r_s\pi}\right)^2 \frac{F_e}{E_e} \tag{9.71}$$

$$P_n = \begin{cases} \left(0.66^\lambda F_e A_s\right), & \text{if } \lambda \le 2.25 \\ \left(\dfrac{0.88 F_e A_s}{\lambda}\right), & \text{if } \lambda > 2.25 \end{cases} \tag{9.72}$$

$$P_e = \frac{A_s F_e}{\lambda} \tag{9.73}$$

$$M_n = \begin{cases} M_{ps}, & \text{if } \dfrac{D}{t} < 2\sqrt{\dfrac{E}{F_y}} \\[3mm] M_{yc}, & \text{if } 2\sqrt{\dfrac{E}{F_y}} \le \dfrac{D}{t} \le 8.8\sqrt{\dfrac{E}{F_y}} \end{cases} \tag{9.74}$$

Requirements from the ASHTO LRFD GS for combined axial compression and flexure are summarized in Equations 9.75 through 9.87. P_n is given by Equation 9.72. M_n is given by Equation 9.87. Presumably, when Extreme Event limit states are being analyzed, the resistance factors on P_n and M_n should both be equal to 1.0 in the equations. This method underestimates flexural resistance by, on average, 25 percent for D/t ratios up to $0.14E/F_y$. Engineers may wish to incorporate this into estimates of maximum load delivered to elements intended to remain elastic during extreme events. Equation 9.78 for P_{ro} is simply Equation 9.72 for P_n using an appropriate resistance factor, along with $\lambda = 0$.

Three methods are available for the determination of M_n. A closed-form method based on exact geometry requires recursive solution of Equation 9.80 for β, the angle (in radians) subtended by the neutral axis chord to the tube center, and subsequent solution for additional parameters to determine M_n as given by Equation 9.87.

$$\frac{P_u}{\phi P_n} + B\left(\frac{M_u}{\phi M_n}\right) \le 1.0 \tag{9.75}$$

$$\frac{M_u}{\phi M_n} \le 1.0 \tag{9.76}$$

$$P_{rc} = 0.75 A_c f_c' \tag{9.77}$$

$$P_{ro} = \phi F_e A_s \tag{9.78}$$

$$B = 1 - \frac{P_{rc}}{P_{ro}} \tag{9.79}$$

$$\beta = \frac{A_s F_y + 0.25 D^2 f_c' \left[\sin\left(\dfrac{\beta}{2}\right) - \sin^2\left(\dfrac{\beta}{2}\right) \tan\left(\dfrac{\beta}{4}\right) \right]}{0.125 D^2 f_c' + D t F_y} \tag{9.80}$$

$$C_r = F_y \beta \frac{Dt}{2} \tag{9.81}$$

$$b_c = D\sin\left(\frac{\beta}{2}\right) \tag{9.82}$$

$$a = \frac{b_c}{2}\tan\left(\frac{\beta}{4}\right) \tag{9.83}$$

$$C_r' = f_c'\left[\frac{\beta D^2}{8} - \frac{b_c}{2}\left(\frac{D}{2} - a\right)\right] \tag{9.84}$$

$$e = b_c\left(\frac{1}{2\pi - \beta} + \frac{1}{\beta}\right) \tag{9.85}$$

$$e' = b_c\left(\frac{1}{2\pi - \beta} + \frac{b_c^2}{1.5\beta D^2 - 6b_c\left(0.5D - a\right)}\right) \tag{9.86}$$

$$\phi M_n = \phi_f\left[C_r e + C_r'e'\right] \tag{9.87}$$

Another closed-form solution for M_n, based on approximate geometry, is simpler and given by Equations 9.88 and 9.89. The section modulus, Z, of the tube alone appears in the equations. This alternative closed-form solution results in lower estimates for M_n compared to those obtained using Equation 9.87. Hence, the engineer may wish to adjust expected moments for capacity design further when using this method to estimate flexural resistance.

$$\phi M_n = \phi_f\left[\left(Z - 2th_n^2\right)F_y + \left(\frac{2}{3}\left(0.5D - t\right)^3 - \left(0.5D - t\right)h_n^2\right)f_c'\right] \tag{9.88}$$

$$h_n = \frac{A_c f_c'}{2Df_c' + 4t\left(2F_y - f_c'\right)} \tag{9.89}$$

A third method for determining axial compression interaction with flexure is to perform a strain compatibility analysis using appropriate constitutive models for the various materials.

9.9.3 Steel Tube Design without Concrete Fill

With no concrete fill, steel tube wall slenderness, D/t, is limited to $0.45E/F_y$ in the AASHTO LRFD BDS, Section 6.12.2.2.3. Flexural resistance of steel tubes without fill is determined as the lesser of that for the limit states of yielding and local buckling, given by Equations 9.90 and 9.91, respectively. The elastic, S, and plastic, Z, section moduli are given in Equations 9.92 and 9.93 for a tube with outer diameter, D, and inner diameter, $D_i = D - 2t$.

$$M_n = F_y Z \qquad (9.90)$$

$$M_n = \begin{cases} \left(\dfrac{0.021E}{D/t} + F_y \right) S, & \text{if } \dfrac{D}{t} \leq \dfrac{0.31E}{F_y} \\[4mm] \dfrac{0.33E}{D/t} S, & \text{if } \dfrac{D}{t} > \dfrac{0.31E}{F_y} \end{cases} \qquad (9.91)$$

$$S = \frac{\pi \left(D^4 - D_i^4 \right)}{32D} \qquad (9.92)$$

$$Z = \frac{D^3 - D_i^3}{6} \qquad (9.93)$$

The shear resistance of steel tubes without fill is given by Equation 9.94 from Section 6.12.1.2.3 of the AASHTO LRFD BDS. The shear length, L_v, is the distance between points of zero and maximum shear in the member under question. Should a point of zero shear not exist, then L_v is taken as the full member length.

$$V_n = 0.5 F_{cr} A_g \qquad (9.94)$$

$$F_{cr} = \text{Max} \begin{cases} \dfrac{1.60E}{\sqrt{\dfrac{L_v}{D}} \left(\dfrac{D}{t} \right)^{5/4}} \leq 0.58 F_y \\[6mm] \dfrac{0.78E}{\left(\dfrac{D}{t} \right)^{3/2}} \leq 0.58 F_y \end{cases} \qquad (9.95)$$

The resistance factor for shear, $\phi_v = 1.00$. The resistance factor for axial compression, $\phi_c = 0.95$, and for axial compression combined with flexure, $\phi_c = 0.90$.

9.9.4 CFST Design for Extreme Event Limit States

For analysis and design of CFST at the Extreme Event limit state for earthquake, ice, and collision loading, a model for inelastic behavior is required. The LRFD BDS is relatively silent on the subject of appropriate models for such analyses.

The Port of Los Angeles (POLA, 2010) and the Port of Long Beach (POLB, 2015) each provide estimated strain limits for both hollow tube and CFST used in wharves. The Washington State Department of Transportation (WSDOT, 2020) also provides useful information for CFST elements.

POLA and POLB each specify three levels of earthquake ground motion for the design of wharves, with corresponding strain limits for each level. The three levels of ground motion correspond to:

- Operating Level Earthquake (OLE) – 50% probability of exceedance in 50 years (72-year mean recurrence interval)
- Contingency Level Earthquake (CLE) – 10% probability of exceedance in 50 years (475-year mean recurrence interval)
- Design Earthquake (DE) – 2% probability of exceedance in 50 years (2,475-year mean recurrence interval). The MRI (mean recurrence interval) for the DE is not explicitly stated, but reference is made to ASCE 7.

Strain limits for both POLA and POLB are summarized in Table 9.2. In-ground plastic hinge strain limits are based on depth to fixity of no more than ten pile diameters. Strain limits for pipe piles at the pile top are based on the use of a concrete plug with dowels into the cap. However, these strain limits do not account for the possibility of local buckling of the tube wall and are likely to only be valid for extremely low D/t values. Equation 9.96, from recent research (Harn et al., 2019), provides strain criteria which capture potential local buckling and are recommended over Table 9.2 values for DBE (design basis earthquake) evaluations.

$$\varepsilon_s \leq 10\left(\frac{D}{t}\right)^{-2} \tag{9.96}$$

Parameters in Table 9.2 are defined as follows.

- ε_c = extreme compressive strain in concrete
- ε_{sd} = extreme tensile strain in dowel reinforcement
- ε_s = steel shell extreme fiber strain
- ε_{smd} = strain at peak stress for dowel reinforcement
- ε_p = extreme tensile strain in prestressing strand
- ρ_s = confining steel volumetric ratio

TABLE 9.2
POLA/POLB Steel Tube Strain Limits

Pile	OLE	CLE	DBE
Concrete (pile top)	$\varepsilon_c \leq 0.005$ $\varepsilon_{sd} \leq 0.015$	$\varepsilon_c \leq (0.005+1.1\rho_s) \leq 0.025$ $\varepsilon_{sd} \leq 0.6\varepsilon_{smd} \leq 0.060$	No limit on ε_c $\varepsilon_{sd} \leq 0.8\varepsilon_{smd} \leq 0.080$
Concrete (in-ground)	$\varepsilon_c \leq 0.005$ $\varepsilon_p \leq 0.015$	$\varepsilon_c \leq (0.005+1.1\rho_s) \leq 0.008$ $\varepsilon_p \leq 0.025$	$\varepsilon_c \leq (0.005+1.1\rho_s) \leq 0.025$ $\varepsilon_p \leq 0.035$
Pipe pile (pile top)	$\varepsilon_c \leq 0.010$ $\varepsilon_{sd} \leq 0.015$	$\varepsilon_c \leq 0.025$ $\varepsilon_{sd} \leq 0.6\varepsilon_{smd} \leq 0.060$	No limit on ε_c $\varepsilon_{sd} \leq 0.8\varepsilon_{smd} \leq 0.080$
Pipe pile (in-ground)	$\varepsilon_s \leq 0.010$	$\varepsilon_s \leq 0.025$	$\varepsilon_s \leq 0.035$
CFST (in-ground)	$\varepsilon_s \leq 0.010$	$\varepsilon_s \leq 0.035$	$\varepsilon_s \leq 0.050$

The WSDOT guidance permits slightly relaxed criteria, compared to LRFD BDS criteria. The slenderness limit for local buckling effects is given by Equation 9.97 for elastic elements or Equation 9.98 for plastic, ductile elements. WSDOT also permits the calculation of shear resistance to include the contribution of concrete fill as shown in Equation 9.99. The coefficient, g_4, is equal to 1.0 under the assumption of fully composite behavior. A reduced value may be required if such conditions are not present.

$$\frac{D}{t} \le 0.22 \frac{E}{F_{yst}} \tag{9.97}$$

$$\frac{D}{t} \le 0.15 \frac{E}{F_{yst}} \tag{9.98}$$

$$\phi V_n = \varphi(g_4)\left[0.6 f_y A_s + 0.0316(3) A_c \sqrt{f_c'}\right] \tag{9.99}$$

9.10 TWO-WAY SHEAR

In addition to requirements for one-way (beam) shear in reinforced concrete, pile caps and similar elements must be assessed for two-way (formerly referred to as "punching") shear. Such checks are typically made at the column and at the most heavily loaded pile. Equation 9.100 defines the design shear resistance for two-way behavior in elements without transverse reinforcement and comprised of normal-weight concrete. The parameter, β_c, is the ratio of the long side to the short side of the area through which the load is transmitted. The perimeter, b_o, surrounds the loaded area at $d_v/2$ from each edge of the loaded area. Recall that the shear depth, d_v, is the distance between the resultant cross-sectional compressive and tensile forces, but must not be less than either $0.72h$ or $0.90d_e$. Since the two-way resistance is that of the pile cap, then d_v is the variable for the pile cap (not the loaded area (column or pile)).

$$\phi V_n = 0.90\left(0.063 + \frac{0.126}{\beta_c}\right)\sqrt{f_c'}\,(b_o d_v) \le 0.126\sqrt{f_c'}\,(b_o d_v) \tag{9.100}$$

Refer to the AASHTO LRFD BDS, Section 5.12.8.6.3, for a full discussion of two-way shear in reinforced concrete pile caps.

9.11 FATIGUE RELATED ISSUES IN REINFORCED CONCRETE

The AASHTO LRFD BDS explicitly exempts concrete deck slabs from fatigue investigation in Section 5.5.3. This same section does require fatigue investigation for reinforcement bar stress in areas where total unfactored dead load (DC plus DW) compressive stress is less than 1.75 times the Fatigue limit state stress range (Fatigue I, infinite life, with impact of 15% included). Equation 9.101 gives the basic fatigue requirement. Equation 9.102 provides the criteria to be used for straight reinforcing bars as well as for welded wire reinforcement, both without welding in the high-stress region, defined as one-third of the span on each side of the point of maximum moment. For locations with bar welds, Equation 9.103 applies. For regular lap-spliced bars, the fatigue threshold, $(\Delta F)_{TH}$ is taken to be equal to 4.0 ksi.

In the equation for fatigue criteria for reinforcement, the yield stress, f_y, is not to be less than 60 ksi, nor more than 100 ksi, regardless of actual, specified yield strength.

The bar stress, f_{min}, is equal to the minimum live load stress from the Fatigue I limit state combined with either (a) the unfactored DC and DW load effects or (b) the unfactored permanent loads, shrinkage, and creep-induced effects. The stress, f_{min}, is positive if tension, negative if compression.

Section properties for fatigue calculations are to be based on cracked sections whenever the sum of stresses, due to unfactored permanent loads combined with the Fatigue I limit state loading, is tensile and exceeds $0.095(f'_c)^{1/2}$.

$$\gamma\left(\Delta f\right) \le \left(\Delta F\right)_{TH} \tag{9.101}$$

$$\left(\Delta F\right)_{TH} = 26 - 22\frac{f_{min}}{f_y} \tag{9.102}$$

$$\left(\Delta F\right)_{TH} = 18 - 0.36 f_{min} \tag{9.103}$$

9.12   ABUTMENT DESIGN

For stub abutments integral with the concrete deck, the design of abutments is relatively simple. With integral abutments, the superstructure effectively prevents large bending moments in the backwall. However, if the construction sequence requires backfilling prior to superstructure installation, the backwall will still need to be designed to resist the fill as a retaining wall.

The abutment beam of a stub abutment on piles may be designed as a continuous, reinforced concrete beam spanning between piles. It is seldom worth the extra effort of a computer model for this design, and bounding assumptions may be made. One such assumption is the plunging of a single pile, producing a span length for the continuous abutment beam equal to twice the pile spacing. While it is certainly possible to place superstructure girder reactions at constructed locations to determine shears and moments in the continuous abutment beam, it is likely to result in an economical, reasonable solution by assuming worst-case locations of girder reactions for shear and moment.

Initially, the total superstructure reaction at the Service and Strength limit states must be determined and combined with self-weight of the abutment backwall and the abutment beam. One proven method for determining the number of point-bearing piles required to support the loads is to assume an equal distribution of vertical load to the piles, and to specify a number of piles such that the Service limit state axial load produces a uniform compressive pile stress no more than 25% of the yield strength of the pile. For friction piles, the Gates formula or the ENR formula may be used to estimate pile resistance from driving logs. Such piles are typically installed with driving logs obtained and compared with results from load testing of a certain percentage of the production piles. The load test permits the use of a higher resistance factor in determining pile resistance at the Strength limit state.

For non-stub abutments, the design is more complex. For large fill behind abutment walls, it becomes necessary to design a *bona fide* retaining wall system for the construction phase, with final bridge loads also accommodated.

A cross-section for a typical stub-type, integral abutment is shown in Figure 9.3. A more complex, non-stub type, fixed abutment is shown in cross section in Figure 9.4.

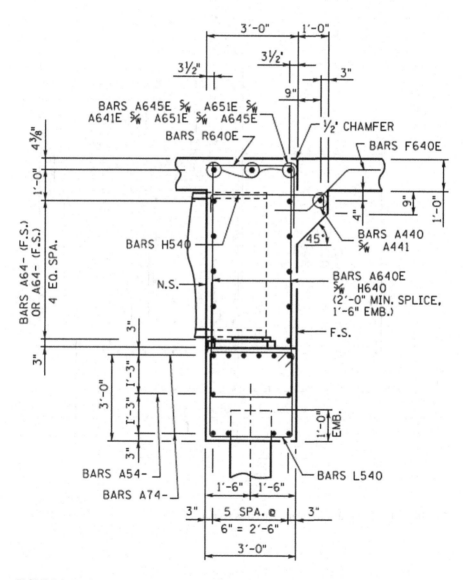

FIGURE 9.3   Cross section of a typical stub-type integral abutment.

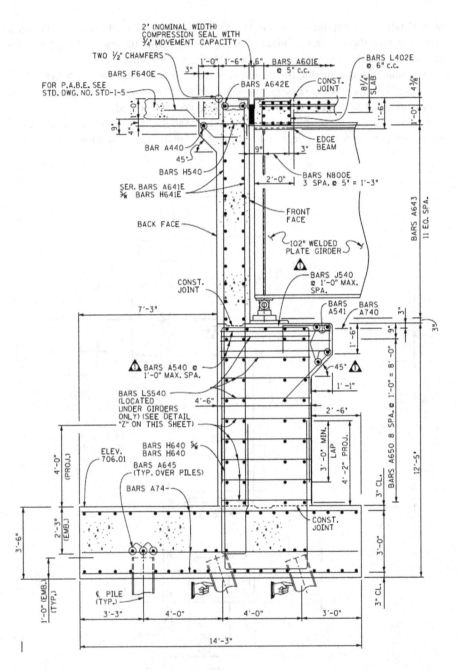

FIGURE 9.4 Non-stub fixed abutment cross section.

9.13 SOLVED PROBLEMS

Problem 9.1

A three-column bent is shown in Figure P9.1. Concrete strengths are, $f'_c=4$ ksi and $f'_{ce}=5.2$ ksi. Steel yield stress values are $f_y=60$ ksi and $f_{ye}=68$ ksi.

Columns are 48 inches in diameter with 30 #9 longitudinal bars and #5 hoops at 5 inches on centers. A 706 reinforcement with reduced ultimate strain $\varepsilon_{su}=0.09$ is used for the hoops and the longitudinal bars. Columns are designed for full fixity at both top and bottom of the column.

Girder reactions for the conditions under consideration include one-half of the live load and are 443 kips for each of the exterior girders, and 509 kips for each of the interior girders.

The total cap length is 45 feet.

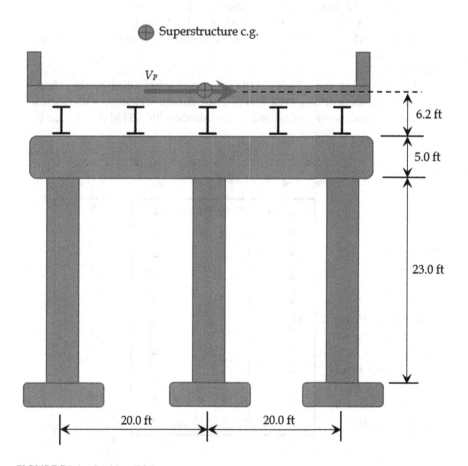

FIGURE P9.1   Problem P9.1.

Determine the displacement capacity of the bent using:

a) AASHTO LRFD GS equations for implicit displacement capacity for Seismic Design Category B.

b) AASHTO LRFD GS equations for implicit displacement capacity for Seismic Design Category C.

c) A simplified pushover analysis, given that the plastic shear, V_P, for the bent has been determined to be 993 kips.

Problem 9.2

A pier cap is shown in Figure P9.2. $f'_c=4$ ksi and $f_y=60$ ksi. Assess the adequacy of the proposed cross section for two cases:

a) Positive moment (tension in the bottom of the cap)
   $M_u=3,720$ ft-k, $V_u=56$ kips (Strength I limit state)
   $M_u=1,950$ ft-k, $V_u=30$ kips (Service I limit state)

b) Negative moment (tension in the top of the cap)
   $M_u=-2,700$ ft-k, $V_u=180$ kips (Strength I limit state)
   $M_u=-1,680$ ft-k, $V_u=112$ kips (Service I limit state)

Use the AASHTO equations and hand calculations. Verify the results using Response 2000.

Problem 9.3

The pile bent shown in Figure P9.3 is used as an intermediate pier in a short-span continuous bridge. The CFST piles are 20 inch in diameter with 5/8 inch in wall thickness and fabricated in accordance with ASTM A53 Grade B,

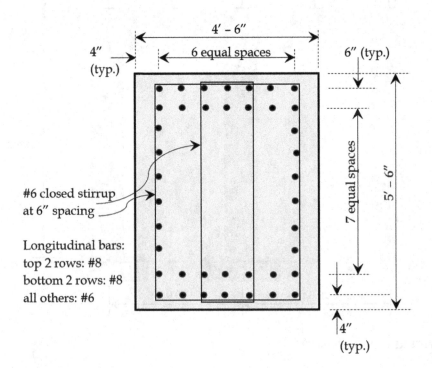

FIGURE P9.2 Problem P9.2.

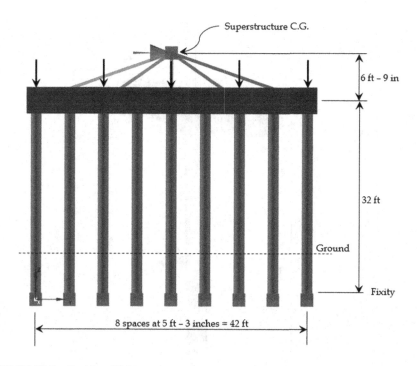

FIGURE P9.3 Problem P9.3.

$F_y=35$ ksi. Concrete core has 28-day compressive strength, $f'_c=3$ ksi. For the conditions shown, each of the five girder-bearing locations carries a vertical reaction of 117 kips. The core is reinforced with six #8 bars.

The cap is 4 ft by 4 ft in cross section. The lateral load shown represents an incremental load to be applied at the superstructure center of gravity in a pushover analysis.

Estimate the lateral load resistance of the bent using one of the methods in AASHTO for the calculation of CFST flexural resistance.

Problem 9.4

For the pier column shown in Figure P9.4, the actions at the column base at the Strength limit state are:

$P_u=3,770$ kips

$V_{uT}=96$ kips $M_{uL}=13,234$ ft-kips

$V_{uL}=67$ kips $M_{uT}=5,915$ ft-kips

The unsupported column height is 67 feet. The column is 6 ft in diameter with 44 #10 bars. Concrete f'_c is equal to 3 ksi. Grade 75 longitudinal bars are used. Grade 60, #5 hoops at 18 inches are used for the full height. Clear cover is 2 inches. Determine the required resistance, M_u, and the design resistance, ϕM_n. Include slenderness effects.

Problem 9.5

For the concrete girder option of the Project Bridge, the pier cap shear and moment for one particular Strength limit state load combination are

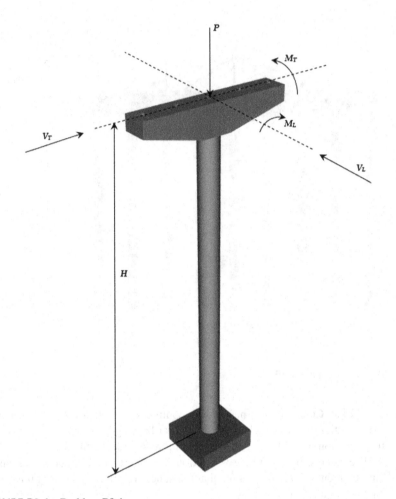

FIGURE P9.4 Problem P9.4

provided below. The cap cross section is shown in Figure P9.5. Dimensions shown are inches. Measured from the top surface, the top four bar layers are at 4 inches, 8 inches, 17 inches, and 26 inches. The specified 28-day concrete compressive strength is $f'_c = 4$ ksi. Steel bar yield stress, $f_y = 60$ ksi. The transverse reinforcement shown is equivalent to four legs of #5 closed stirrups spaced at 6 inches apart. Determine the design shear resistance, ϕV_n, and flexural resistance, ϕM_n, at the section under consideration. Compare these resistance values with the required resistance values.

$M_u = -1,462$ ft-kips (tension in the top of the cap)

$V_u = 311$ kips

Problem 9.6

For the concrete girder option of the Project Bridge, the column demands for one particular Strength limit state load combination are given below.

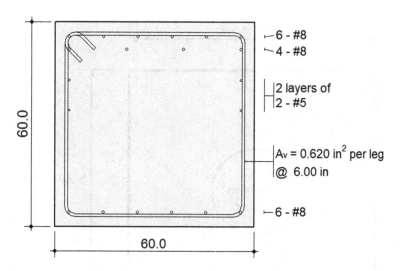

60.0

60.0

6 - #8
4 - #8

2 layers of
2 - #5

Av = 0.620 in^2 per leg
@ 6.00 in

6 - #8

FIGURE P9.5 Problem 9.5

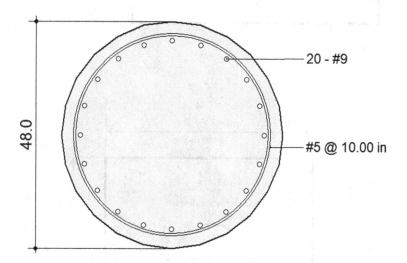

48.0

20 - #9

#5 @ 10.00 in

FIGURE P9.6 Problem 9.6.

The column cross section is shown in Figure P9.6. Dimensions shown are inches. The diameter of the bar circle is 40 inches. Determine the design resistance values, ϕM_n, ϕV_n, and ϕP_n, and compare with the required values.

$P_u = 1,231$ kips
$M_{ux} = 1,054$ ft-kips
$M_{uz} = 345$ ft-kips
$V_{uz} = 44$ kips
$V_{ux} = 30$ kips

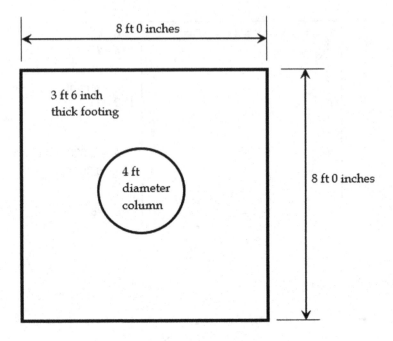

FIGURE P9.7   Problem 9.7.

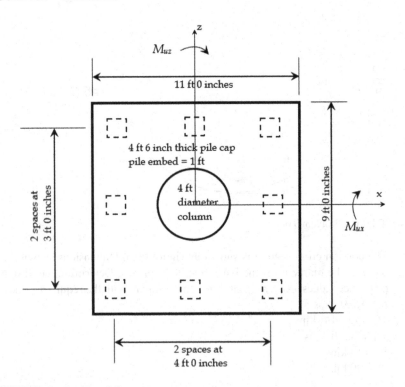

FIGURE P9.8   Problem 9.8.

Problem 9.7

For the column base loads given in Problem 9.6, determine the maximum pressure using the spread footing dimensions shown in Figure P9.7. The footing has 2 feet of soil fill over the top of the footing. Soil unit weight is 120 pounds per cubic foot.

Problem 9.8

For the column base loads given in Problem 9.6, determine the maximum pile reaction using the pile cap dimensions shown in Figure P9.8. The pile cap has 2 feet of soil fill over the top of the footing. Soil unit weight is 120 pounds per cubic foot. Determine the required blow count during pile driving using the Gates formula and the ENR formula with appropriate resistance factors. The pile hammer weighs 7,930 pounds and has a stroke of 127 inches.

PROBLEM 9.1	TEH	1/5

a) SDC "B" implicit equations

$$B_o = 4 \text{ ft}$$

$$H_o = 23 \text{ ft}$$

$$\Lambda = 2$$

$$x = \frac{2(4)}{23} = 0.348$$

$$\Delta_c = 0.12(23)[-1.27 \ln(.348) - 0.32]$$

$$= 0.1225(23) = \underline{\underline{2.82''}}$$

b) SDC "C" implicit equations

$$\Delta_c = 0.12(23)[-2.32 \ln(.348) - 1.22]$$

$$= 0.148(23) = \underline{\underline{3.39''}}$$

c) Pushover with $V_p = 993$ kips

993^k

mid-height of cols

V_1 V_2 V_3

P_1 P_2 P_3

6.2 + 5
+ 23/2
= 22.7'

PROBLEM 9.1	TEH	2/5

$W_{cap} = 45 \times 5 \times 5 \times 0.15 = 169 \text{ kips}$

$W_{col} = \pi (4)^2 / 4 \times 0.15 \times 23' = 43 \text{ kips}$
$$(each)$$

Assume loads (vertical) are equally distributed to the 3 columns. Refinement to these calculations could include a frame analysis to establish the true distribution of vertical loads.

$W_{SS} = 2 \times 443 + 3 \times 509 = 2,413^k$

$W_{TOT} = 169 + 3 \times 43 + 2,413$

$$= 2,711 \text{ kips}$$

$$\div 3 = 904 \text{ kips/column}$$

Overturning due to V_p:

$$993 (22.7) = \Delta P (\Delta_0)$$

$$\Rightarrow \Delta P = 563 \text{ kips}$$

$P_1 = 904 - 563 = 341 \text{ kips}$

$P_2 = 904 \text{ kips}$

$P_3 = 904 + 563 = 1,467 \text{ kips}$

PROBLEM 9.1	TEH	3/5

The most heavily loaded column, column 3, will be the least ductile. Use $P = P_3 = 1,467$ kips

$$c/D \cong 0.20 + \frac{0.65(1,467)}{5.2(1,810)}$$

$$= 0.301$$

$$\Rightarrow C = 0.301 \times 48 = 14.46''$$

$$S = 5'' \qquad S' = 5 - 5/8 = 4.375''$$

$$d_s = 48 - 2(2.5) - 2(5/8 \times 1/2)$$

$$\Rightarrow d_s = 42.375''$$

$$A_c = \pi(42.375)^2/4 = 1,410 \text{ in}^2$$

$$\rho_{cc} = \frac{30 \times 1.00}{1,410} = 0.0213$$

$$\rho_v = \frac{4 \times 0.31}{42.375 \times 5} = 0.00585$$

$$k_e = \frac{\left(1 - \frac{4.375}{2 \times 42.375}\right)^2}{1 - 0.0213} = 0.919$$

$$f'_\ell = \frac{1}{2} \times 0.919 \times 0.00585 \times 60$$

$$\Rightarrow f'_\ell = 0.161 \text{ ksi}$$

$$f'_\ell / f'_{co} = 0.161 / 4 = 0.0403$$

PROBLEM 9.1	TEH	4/5

$$f_{cc} = 4\left[-1.254 + 2.254\sqrt{1 + 7.94 \times .0403} - 2 \times .0403\right]$$

$$= 4(1.255) = 5.021 \text{ ksi}$$

$$\varepsilon_{cu} = 0.004 + \frac{1.4(.00585)(60)(0.09)}{5.021}$$

$$\Rightarrow \varepsilon_{cu} = 0.01281$$

$$\varepsilon_y = 60/29{,}000 = 0.00207$$

$$\phi_y = 2.25(0.00207)/48$$

$$\Rightarrow \phi_y = 0.0000970 \text{ in}^{-1}$$

$$\phi_u = \text{Min} \begin{cases} 0.01281/14.46 = 0.000886 \\ 0.09/(48-14.46) = 0.00268 \end{cases}$$

$$\Rightarrow \phi_u = 0.000886 \text{ in}^{-1}$$

$$L_c = \frac{23 \times 12}{2} = 138''$$

$$L_{sp} = 0.15 \times 68 \times 1.128 = 11.5''$$

$$0.08 L_c = 0.08 \times 138 = 11.0''$$

$$\Rightarrow L_p = 2 \times 11.5 = 23.0''$$

$$\Delta_y = 2\,(1/3)(0.0000970)(138+11.5)^2$$

$$\Rightarrow \underline{\Delta_y = 1.44''}$$

PROBLEM 9.1	TEH	5/5

$$\Delta_p = (0.000789)(23.0)(138) = 2.504''$$
$$\text{between mid-ht.}$$
$$\text{& each end}$$

$$\Delta_p = 2 \times 2.504 = 5.01'', \text{ total}$$

$$\Delta_u = 1.44 + 5.01$$

$$\Rightarrow \underline{\Delta_u = 6.45''}$$

PROBLEM 9.2	TEH	1/5

Response 2000 model:
$$\text{"CEE4380-P09.02"}$$

a) Positive Moment

Strength I : $M/V = 3,720/56$
$$= 66.43 \text{ ft}$$

d_v = distance between resultant tensile & compressive forces

$d_v = 21.44 + 28.65 = 50.09''$

$V_u \times d_v = 56 \times 50.09 /12 = 234 \text{ ft·K}$

Take $\varepsilon_s = 0.006$

$$\Rightarrow \beta = \frac{4.8}{1 + 4.5} = 0.873$$

$$\theta = 29 + 21 = 50°$$

$V_c = 0.0316 (0.873)\sqrt{4} (54)(50.09)$

$$\Rightarrow V_c = 149.2 \text{ KIPS}$$

$V_s = (4 \times 0.44)(60)(50.09)/6$

$$\Rightarrow V_s = 881.6 \text{ KIPS}$$

$V_n = 149.2 + 881.6 = 1,031 \text{ KIPS}$

$(V_n)_{MAX} = 0.25(4)(54)(50.09)$

$$= 2,705 \text{ KIPS}$$

$$\Rightarrow \phi V_n = 0.9 \times 1,031 = 928 \text{ KIPS}$$
$$> V_u = 56 \text{ KIPS}, \text{ OK BY}$$
$$\text{AASHTO Equations}$$

PROBLEM 9.2	TEH	2/5

Assume all bars in lower half of the cross-section yield and verify.

$A_s = 14 \times .79 + 6 \times 0.44 = 13.70 \, in^2$

$$\bar{y}_s = \frac{(7 \times .79)(4+10) + 2(.44)(16.57 + 23.14 + 29.7)}{13.70}$$

$\bar{y}_s = 10.11''$

$d = 66 - 10.11 = 55.89''$

$$\frac{a}{2} = \frac{60 \times 13.70}{2 \times .85 \times 4 \times 54} = 2.24''$$

Assume tension-control & verify.

$\phi M_n = .9 \times 60 \times 13.70 (55.89 - 2.24) / 12$

$\Rightarrow \phi M_n = 0.9 \times 3,675 = 3,308 \, ft \cdot k$

$c = \frac{a}{.85} = \frac{2 \times 2.24}{0.85}$

$\Rightarrow c = 5.27''$

$\varepsilon_t = 0.003 \left(\frac{62 - 5.27}{5.27} \right)$

$\Rightarrow \varepsilon_t = 0.0323$

$> \varepsilon_{tl} = 0.005$

$\Rightarrow$ tension-control verified

$\varepsilon_2 = 0.0323 \left(\frac{62 - 5.27 - 29.7}{62 - 5.27} \right) = 0.0154$

PROBLEM 9.2	TEH	3/5

$\varepsilon_2 = 0.0154 > \varepsilon_y = \frac{60}{29,000} = 0.00207$

$\Rightarrow$ yield assumption verified

$\phi M_n = .9 \times 3,675 = 3,308 \text{ ft} \cdot \text{k}$
$\qquad \text{(hand calc.)}$

$\phi M_n = .9 \times 3,603 = 3,243 \text{ ft} \cdot \text{k}$
$\qquad \text{(R2000)}$

Either way, $\phi M_n < M_u = 3,720$
$\qquad \Rightarrow \underline{\text{No Good}}$
$\qquad \underline{(M^+, \text{Strength I})}$

From R2000, when $M_u = 1,950 \text{ ft} \cdot \text{k}$
$\qquad\qquad V_u = 30 \text{ kips}$

$\Rightarrow f_{ss} = 34.6 \text{ ksi}$
$d_c = 4''$

$\beta_s = 1 + \frac{4}{0.7(66-4)} = 1.092$

$\gamma_e = 1.00$

$S \leq \frac{700(1.00)}{1.092(34.6)} - 2(4) = 10.5''$

$S_{actual} = (54-8)/6 = 7.67'' < 10.5''$

$\underline{\text{ok } (M^+, \text{Service})}$

PROBLEM 9.2	TEH	4/5

Check side steel

$$A_{sk} \geqslant 0.012(66-30) = 0.432 \text{ in}^2/\text{ft}$$
$$\text{each side}$$

Side bar spacing $= (66-20)/7$
$$= 6.57''$$

$$A_{sk} = 0.44 \left(\frac{12}{6.57}\right) = 0.803 \text{ in}^2/\text{ft}$$
$$> 0.432 \text{ in}^2/\text{ft}$$

$$S_{max} = 66/6 = 11'' \quad, \quad S = 6.57'' < 11''$$

<u>Side bars OK</u>

b) Negative Moment

Strength I : $M/V = 2,700/180$
$$= 15.0 \text{ ft}$$

R2000 : $d_v = 22.79 + 24.08$
$$= 46.87''$$

$V_u \times d_v = 180 \times 46.87/12 = 703 \text{ ft·k}$

R2000 : $M_n = 2,945 \text{ ft·k}$

$\phi M_n = .9 \times 2,945 = 2,650 \text{ ft·k}$

$< M_u = 2,700 \text{ ft·L}$

$\Rightarrow$ No Good
<u>(M⁻, Strength)</u>

Service $M_u = -1,680\ ft\cdot k$

$V_u = 112\ k$

$\Rightarrow$ R2000 $f_{ss} = 35.0\ ksi$

$S \leq \dfrac{700(1.00)}{1.092(35.0)} - 2(4)$

$= 10.3''$

$S_{actual} = 7.67'' < 10.3''$

$\underline{ok\ (M^-,\ Service)}$

PROBLEM 9.3	TEH	1/5

CFST   20" o.d. $5/8"$ thick
            18.75" i.d.   D/t = 32.0

$A_c = \pi(18.75)^2/4 = 276\ in^2$

$A_{st} = \pi(20^2 - 18.75^2)/4 = 38.0\ in^2$

$\dfrac{A_{st}}{A_{tot}} = \dfrac{38.0}{314} = 0.121 > 0.040,\ OK$

Use BDS 6.9.5, 6.12.3.2.2, and
       GS 7.6.

For the tube alone:

$$Z = (20^3 - 18.75^3)/6$$

$$= 234.7\ in^3$$

$$S = \dfrac{\pi(20^4 - 18.75^4)}{32(20)} = 178.7\ in^3$$

$A_r = 6 \times 0.79 = 4.74\ in^2$

$n = 29,000/(1820\sqrt{3}) = 9.2$

$F_e = 35 + 60\left(\dfrac{4.74}{38.0}\right) + 0.85(3)\left(\dfrac{276}{38.0}\right)$

$\Rightarrow F_e = 61.0\ ksi$

$E_e = 29,000\left[1 + \left(\dfrac{40}{9.2}\right)\left(\dfrac{276}{38.0}\right)\right]$

$\Rightarrow E_e = 38,158\ ksi$

Take $L = 32' - \frac{1}{2}(4') = 30' = 360"$

| PROBLEM 9.3 | TEH | 2/5 |

Assume the bent is braced by the deck for out-of-plane buckling; Assume rigid-frame behavior for in-plane buckling.

$$\Rightarrow K = 1.2 \quad (BDS\ 4.6.2.5)$$

$$I_s = \pi(20^4 - 18.75^4)/64 = 1,787\ in^4$$

$$r_s = (1,787/38.0)^{1/2} = 6.86\ in$$

$$\lambda = \left(\frac{1.2 \times 360}{6.86\ \pi}\right)^2 \left(\frac{61.0}{38,158}\right) = 0.643$$

Since $\lambda < 2.25$:

$$P_n = 0.66^{0.643}(61.0)(38.0)$$

$$\Rightarrow P_n = 1,775\ kips$$

$$\underline{\phi P_n = .9\ P_n = 1,597\ kips}$$

For moment magnification $(BDS\ 4.5.3.2.2)$

$$P_e = 38.0(61.0)/0.643$$

$$\Rightarrow P_e = 3,605\ kips\ per\ pile$$

$$P_u = 117\left(\frac{5}{9}\right) + (4 \times 4 \times 45 \times 0.15 \times 1.25)/9$$

$$\Rightarrow P_u = 80\ kips\ per\ pile$$

PROBLEM 9.3	TEH	3/5

Moment magnifier, δ

$$\delta = \frac{1}{1 - \frac{80}{3605}} = 1.023$$

$P_{rc} = 0.75 \times 276 \times 3 = 621^K$

$P_{ro} = 0.90 \times 61.0 \times 38.0 = 2,086^K$

$B = 1 - \frac{621}{2,086} = 0.702$

From trial & error, $\beta = 2.4326$ rad

$\Rightarrow C_r = 35(2.4326)(20 \times .625)/2$

$\Rightarrow C_r = 532$ kips

$b_e = 20 \sin(2.4326/2) = 18.76$ in

$a = (18.76/2) \tan(2.4326/4) = 6.53$ in.

$C'_r = 3 \left[2.4326 \frac{(20)^2}{8} - \frac{18.76}{2}(10 - 6.53) \right]$

$\Rightarrow C'_r = 267$ kips

$e = 18.76 \left(\frac{1}{2\pi - 2.4326} + \frac{1}{2.4326} \right)$

$\Rightarrow e = 12.58$ in.

$e' = 18.76 (0.2597 + 0.3292)$

$\Rightarrow e' = 11.05$ in.

| PROBLEM 9.3 | TEH | 4/5 |

$$\phi M_n = 0.9 \left(532 \times 12.58 + 267 \times 11.05 \right)$$

$$= 0.9 \left(9,643 \ \text{in-k} \right)$$

$$= 8,679 \quad \text{in-k}$$

$$= 723 \ \text{ft-k} \quad \text{per pile}$$

$$M_u = \frac{V_{u\text{-TOTAL}}}{9 \ \text{piles}} \times \left(\tfrac{1}{2} \times 30 + 4 + 4.75 \right)(1.023)$$

$$\Rightarrow M_u = 2.7 \ V_{u\text{-TOTAL}}$$

(Assumed contraflexure at mid-height)

Estimate increase in axial load
due to overturning to be minimal,
say 10 kips, and verify:

$$P_u = 80 + 10 = 90^k$$

$$\frac{90}{1,597} + 0.702 \left(\frac{2.7 \ V_{u\text{-TOTAL}}}{723} \right) \le 1.0$$

$$\Rightarrow V_{u\text{-TOTAL}} \le 360 \ \text{kips}$$

$$\frac{2.7 \ V_{u\text{-TOTAL}}}{723} \le 1.0$$

$$\Rightarrow V_{u\text{-TOTAL}} \le 268 \ \text{kips} \xleftarrow{\ } \text{controls}$$

PROBLEM 9.3	TEH	5/5

Overturning Effect on P_u:

$$\Sigma_i d^2 = 2(5.25^2 + 10.5^2 + 15.75^2 + 21^2)$$

$$= 1,654 \, ft^2$$

$$M_{oT} = 268^K \left(\frac{1}{2} \times 30 + 8.75\right)(1.023)$$

$$= 6,511 \, ft \cdot K$$

$$\Delta P_u = 6,511 \,(21)/1,654 = 83 \, kips$$

$$> 10 \, kips \text{ assumed}:$$

$$P_u = 80 + 83 = 163^K$$

$$\frac{163}{1,597} + 0.702 \left(\frac{2.7 V_{u\text{-TOTAL}}}{723}\right) \leq 1.0$$

$$\Rightarrow V_{u\text{-TOTAL}} \leq 342^K$$

Moment criteria still controls:

$$\underline{V_{u\text{-TOTAL}} \leq 268^K}$$

Shear resistance: $F_{cr} = \frac{.78 \times 29,000}{32^{1.5}} = 112 \, ksi$

$$> .58 \times 35 = 20.3 \, ksi$$

$$\Rightarrow \phi V_n = 1.00 \times 0.5 \times 20.3 \times 38.0$$

$$= 386^K/pile$$

$$> V_u = 268/9 = 30^K/pile, \, OK$$

| PROBLEM 9.4 | TEH | 1/5 |

$$A = \pi(8)^2/4 = 50.26 \ ft^2$$

$$I = \pi(8)^4/64 = 201.1 \ ft^4$$

$$r = \sqrt{\frac{201.1}{50.26}} = 2.00 \ ft$$

In the absence of additional info, take $K = 2.1$ (LRFD-BDS 4.6.2.5)

$$Kl_u/r = 2.1 \times 67/2.00 = 70.35 < 100$$

$\Rightarrow$ Approximate slenderness effect analysis is ok.

$$EI = \frac{E_c I_g}{2.5(1+\beta_d)} \qquad (LRFD\text{-}BDS \ 5.6.4.3)$$

$$\beta_d \cong 0 \qquad E_c = 1,820\sqrt{3} = 3,152 \ ksi$$

$$P_e = \frac{\pi^2(3,152 \times 144)(201.1)}{(2.1 \times 67)^2} = 45,506 \ kips$$

$$\delta_s = \frac{1}{1 - \frac{3,770}{.75 \times 45,506}} = 1.124$$

(LRFD-BDS 4.5.3.2.2b)

For circular column, use resultant moment, M_{ur}.

PROBLEM 9.4	TEH	2/5

1st- order

$$M_{ur} = (13,234^2 + 5,915^2)^{1/2}$$

$$= 14,496 \text{ ft·k}$$

2nd- order

$$M'_{ur} = 1.124 \times 14,496$$

$$= 16,296 \text{ ft·k}$$

$$V_{ur} = (96^2 + 67^2)^{1/2} = 117 \text{ KIPS}$$

$$M/V = 16,296/117 = 139'$$

Response - 2000 file "CEE4380-P09.04"

With $P_u = -3,770$ KIPS

$\quad$ $M/V = 139$ ft

$\quad \Rightarrow M_n = 22,788$ ft·k

Compression - controlled strain limit:

$$\varepsilon_{cl} = 75/29,000 = 0.0026$$

Tension - controlled strain limit:

$$\varepsilon_{tl} = 0.005$$

(LRFD - BDS 5.6.2.1)

PROBLEM 9.4	TEH	3/5

From R-2000 Plots :

$$\varepsilon_t = 0.00394$$

LRFD-BDS 5.5.4.2 :

$$\phi = 0.75 + 0.15 \left(\frac{0.00394 - 0.0026}{0.005 - 0.0026} \right)$$

$$\Rightarrow \phi = 0.834$$

$$\phi M_n = 0.834 \times 22,788$$

$$= 18,999 \text{ ft·k}$$

$$> M_u = 16,296 \text{ ft·k}$$

$$\Rightarrow \underline{\underline{ok}}$$

96.0

96.0 — 44 - #11

96.0 — #6 @ 10.00 in

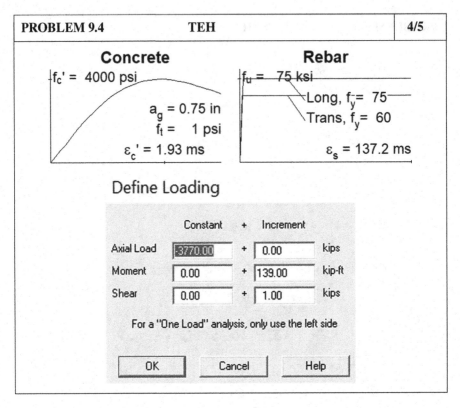

Concrete

$f_c' =$ 4000 psi

$a_g = 0.75$ in
$f_t = 1$ psi
$\varepsilon_c' = 1.93$ ms

Rebar

$f_u = 75$ ksi

Long, $f_y = 75$
Trans, $f_y = 60$

$\varepsilon_s = 137.2$ ms

Define Loading

	Constant	+	Increment	
Axial Load	-3770.00	+	0.00	kips
Moment	0.00	+	139.00	kip-ft
Shear	0.00	+	1.00	kips

For a "One Load" analysis, only use the left side

| OK | Cancel | Help |

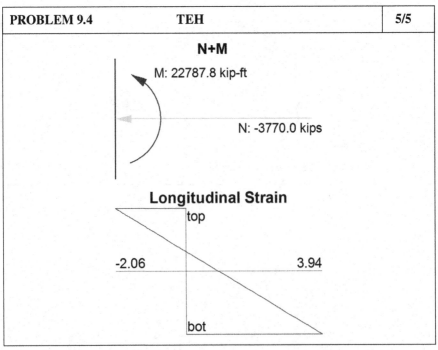

N+M

M: 22787.8 kip-ft

N: -3770.0 kips

Longitudinal Strain

top

-2.06   3.94

bot

PROBLEM 9.5	TEH	1/4

Check Minimum A_v:

$$A_v \geqslant 0.0316 \sqrt{4} \; \frac{(60)(6)}{60} = 0.379 \, in^2$$

$$A_v = 4 \times 0.31 = 1.24 \, in^2 > 0.379$$
$$\Rightarrow OK$$

Determine shear depth, d_v
$$d_e = 0.72 \times 60 = 43.2"$$
(since we don't know d_e yet)

Determine shear stress
$$\mathcal{V}_u = \frac{311}{0.9 \times 60 \times 43.2} = 0.133 \, ksi$$

$$0.125 f_c' = 0.125 \times 4 = 0.50 \, ksi$$

$$0.133 < 0.500 \Rightarrow S_{max} = .8 \times 43.2$$
$$= 34.5"$$

$S_{max} = 24"$ controls
$$S = 6" < 24" \Rightarrow OK$$

$$V_u \times d_v = 311 \times 43.2 / 12 = 1,120 \, ft \cdot k$$
$$M_u = 1,462 \, ft \cdot k > V_u d_v$$
$$\Rightarrow use \; M_u = 1,462 \, ft \cdot k$$

$$A_s = 10 \, \#8 + 4 \, \#5 = 10(0.79) + 4(0.31)$$
$$\Rightarrow A_s = 9.14 \, in^2$$

| PROBLEM 9.5 | TEH | 2/4 |

$$\varepsilon_s = \frac{\frac{1,462 \times 12}{43.2} + 311}{29,000 \times 9.14} = 0.0027$$

$$\beta = \frac{4.8}{1 + 750(.0027)} \implies \beta = 1.58$$

$$\theta = 29 + 3500(.0027) \implies \theta = 38.4°$$

$$V_c = 0.0316 (1.58)\sqrt{4} (60)(43.2)$$
$$\implies V_c = 260 \text{ KIPS}$$

$$V_s = \frac{1.24 \times 60 \times 43.2}{6} \cot 38.4°$$
$$\implies V_s = 676 \text{ KIPS}$$

$$\phi V_n = 0.90 (260 + 676) = 842^k$$
$$> V_u = 311^k$$
$$\implies \underline{\text{Shear OK (by eq'ns)}}$$

However, the M/V ratio for this loading is relatively small:

$$M_u / V_u = \frac{1,462}{311} = 4.7 \text{ ft}$$

(member is 5' deep)

Use Response 2000 with M/V interaction rather than a "standard" flexural analysis.

PROBLEM 9.5	TEH	3/4

From Response 2000:

$$d_v = 43.12'' \text{, very close to } 43.2'' \text{ used previously}$$

$M_n = 1,960 \text{ ft} \cdot k$

$\varepsilon_t = 0.00275 \Rightarrow \text{transition}$

$\varepsilon_{cl} = 0.002 \qquad \varepsilon_{tl} = 0.005$

$\phi = 0.75 + 0.15 \dfrac{(.00275 - .002)}{(.005 - .002)}$

$\Rightarrow \phi = 0.787$

$\phi M_n = 0.787 \times 1,960$

$\qquad = 1,543 \text{ ft} \cdot k$

$\qquad > M_u = 1,462 \text{ ft} \cdot k \text{, } \underline{ok}$

Note: Section Analysis without M/V Interaction gives:

$M_n = 2,342 \text{ ft} \cdot k$

$\varepsilon_t = 0.032$

$\phi = 0.90$

$\phi M_n = 2,108 \text{ ft} \cdot k$

(False margin of safety)

PROBLEM 9.5	TEH	4/4

Response 2000 results from "CEE4380-P09-05.rsp":

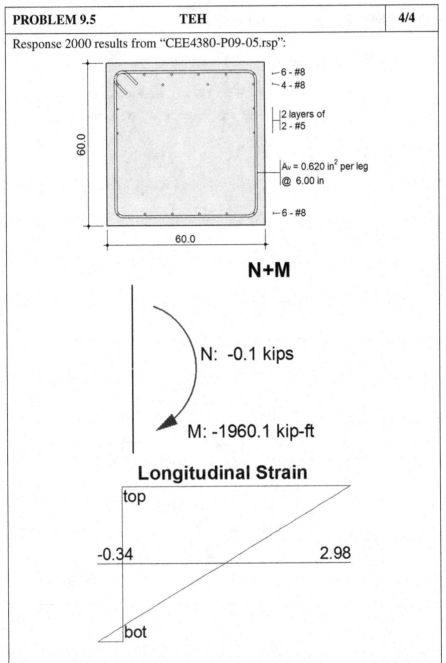

The strain in the concrete at the tension face is 0.00298. Response 2000 shows that the strain in the reinforcement is $\varepsilon t = 0.00275$.

PROBLEM 9.6	TEH	1/2

Circular Column

$$M_{ur} = (1{,}054^2 + 345^2)^{1/2}$$

$$\Rightarrow M_{ur} = 1{,}109 \ ft \cdot k$$

$$V_{ur} = (44^2 + 30^2)^{1/2}$$

$$\Rightarrow V_{ur} = 53.2 \ kips$$

$$M_{ur}/V_{ur} = 1{,}109 / 53.2$$

$$= 20.8 \ ft$$

Response 2000 : (w/ M/V = 20.8')

$$M_n = 2{,}985 \ ft \cdot k$$

$$\varepsilon_t = 0.00301$$

$$\varepsilon_{tl} = 0.005 \ , \ \varepsilon_{cl} = 0.002$$

$$\phi = 0.75 + 0.15 \frac{(.00301 - .002)}{(.005 - .002)}$$

$$\Rightarrow \phi = 0.80$$

$$\phi M_n = 0.80 \times 2{,}985$$

$$= 2{,}389 \ ft \cdot k$$

$$> M_u = 1{,}109 \ ft \cdot k$$

$$\Rightarrow \underline{\underline{ok}}$$

PROBLEM 9.6	TEH	2/2

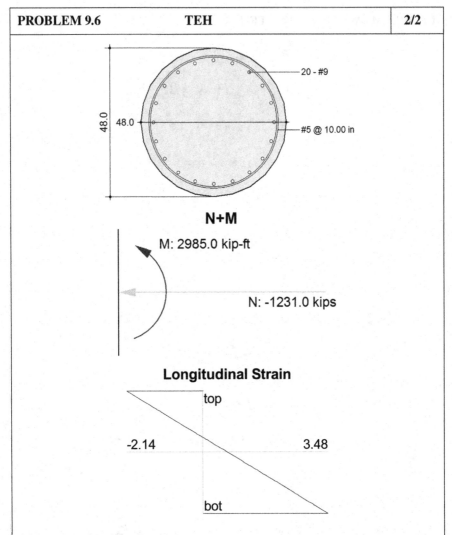

The strain at the extreme tension face is 0.00348. From the response 2000 data, the strain at the steel closest to the tension face is $\varepsilon t = 0.00301$

$$A_{FTG} = 8' \times 8' = 64 \text{ ft}^2$$

$$S_x = S_z = 8(8)^2/6 = 85.33 \text{ ft}^3$$

Soil Weight, $\gamma_{EV} = 1.35$ (Strength)

$$W_{SOIL} = 1.35(2')(12 \times 8^2 - \pi(4)^2/4)$$

$$= 17 \text{ KIPS}$$

Footing Weight, $\gamma_{DC} = 1.25$

$$W_{FTG} = 1.25(.15)(8 \times 8 \times 3.5)$$

$$= 42 \text{ KIPS}$$

$$P_u' = 1,231 + 17 + 42 = 1,290 \text{ KIPS}$$

$$q = \frac{1,290}{64} \pm \frac{1,054}{85.33} \pm \frac{345}{85.33}$$

$$= 20.16 \pm 12.35 \pm 4.04$$

$$\Rightarrow q_{MAX} = 36.6 \text{ KSF}$$

$$q_{MIN} = 3.8 \text{ KSF} > 0$$

$$\Rightarrow \text{Max is correct}$$

$$\Rightarrow q_{max} = 36.6 \text{ KSF}$$

PROBLEM 9.8	TEH	1/2

$$\text{Soil } wt = 1.35(.12)(2')(11 \times 9 - \pi(4)^2/4)$$

$$= 28 \text{ KIPS}$$

$$\text{Footing } wt = 1.25(.15)(11 \times 9 \times 4.5)$$

$$= 84 \text{ KIPS}$$

$$P_u' = 1,231 + 28 + 84 = 1,343 \text{ KIPS}$$

$$N = 8 \text{ piles}$$

$$I_x = 6(3)^2 = 54 \text{ ft}^2$$

$$I_z = 6(4)^2 = 96 \text{ ft}^2$$

$$P_{pile} = \frac{1343}{8} \pm \frac{1,054(3)}{54} \pm \frac{345(4)}{96}$$

$$= 168 \pm 59 \pm 14$$

$$P_{MAX} = 241 \text{ KIPS}$$

$$P_{MIN} = 95 \text{ KIPS} > 0 \text{ (No uplift)}$$

Gate's Formula

$$E_d = 7,930 \times 127/12$$

$$= 83,926 \text{ ft} \cdot \text{lbs}$$

$$\sqrt{E_d} = 289.7$$

PROBLEM 9.8	TEH	2/2

$$\phi R_{ndr} = .4 \left[(1.75)(289.7) \log_{10}(10N_b) - 100 \right]$$

$$\geqslant 241 \ KIPS$$

$$\log_{10}(10 N_b) \geqslant 1.386$$

$$\Rightarrow N_b \geqslant 2.43 \ blows/inch$$

$$\Rightarrow N_b \geqslant 29.2 \ blows/ft$$

<u>ENR Formula</u>

$$E_d = 83.93 \ ft \cdot k$$

$$\phi R_{ndr} = 0.10 \left(\frac{12 \times 83.93}{5 + 0.1} \right)$$

$$\geqslant 241 \ KIPS$$

$$\Rightarrow S \leqslant 0.318 \ inches/blow$$

$$\Rightarrow N \geqslant \frac{1}{0.318} = 3.15 \ \frac{blows}{inch}$$

$$\Rightarrow N \geqslant 37.7 \ blows/ft$$

9.14 EXERCISES

E9.1.

For the pile bent described in Problem 9.3, estimate the nominal flexural resistance using (a) the approximate method in AASHTO and (b) a cross-section analysis program. Also determine the maximum strain which could be used in a pushover analysis based on buckling of the tube wall.

E9.2.

An 8 ft diameter drilled shaft is subjected to the Strength limit state loading given below. The shaft has forty #14 longitudinal bars and #6 hoops spaced at 12 inches on center. Material properties are: $f_y=60$ ksi, $f'_c=4$ ksi. The distance from the outer surface of the shaft to the center of the longitudinal bars is 4.25 inches. Determine whether the shaft satisfies Strength limit state requirements for the loading condition given.

- $P_u=4,675$ kips, axial compression
- $M_u=22,540$ ft-kips, moment
- $V_u=644$ kips, shear

E9.3.

A 14 inch square prestressed concrete pile embedded 12 inches into a pile cap constructed from 3,000 psi concrete receives a Strength limit state load of 220 kips. Determine the minimum total thickness of the pile cap required based on two-way shear requirements. The pile in question is at the corner of the pile cap. The distance from the edge of the cap to the center of the pile is 1 ft 6 inches.

E9.4.

A spread footing supports a column with Strength limit state axial load, $P_u=2,300$ kips, at the base of the column. The footing is 12 ft $\times$ 12 ft in plan and is 5 feet thick. Moments at the Strength limit state acting in unison with the given axial load are $M_{ux}=7,475$ ft-kips and $M_{uz}=6,400$ ft-kips. Determine the maximum soil pressure exerted by the footing (a) assuming a uniform pressure distribution and (b) assuming a linearly varying pressure distribution. Check eccentricity limits as well.

E9.5.

Determine the minimum and maximum pile axial loads. The factored loads at the base of the column are:

- $P_u=800$ kips
- $M_{ux}=2,000$ ft-kips
- $M_{uz}=1,800$ ft-kips

The 11 ft $\times$ 11 ft $\times$ 5.5 ft pile cap is shown in Figure E9.5. There is no pile underneath the circular column in the center of the pile cap.

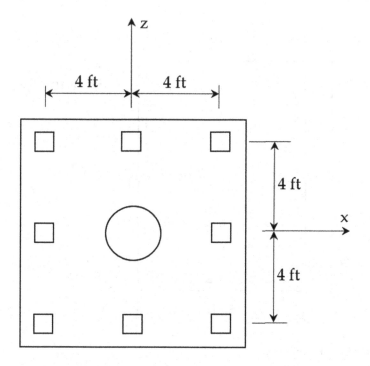

FIGURE E9.5 Exercise E9.5.

E9.6.

A rectangular pier column ($f'_c = 4$ ksi, $f_y = 60$ ksi) is subjected to the following loads at the Strength limit state:

- $P_u = 1,877$ kips
- $M_{ux} = 3,250$ ft-kips
- $M_{uz} = 5,126$ ft-kips

The cross section is shown in Figure E9.6. Use the biaxial flexure approximate solution to determine if the column satisfies the Strength limit state resistance requirements.

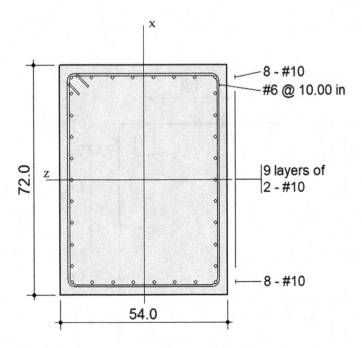

FIGURE E9.6   Exercise E9.6.

10 Seismic Design of Bridges

This chapter is intended to provide a discussion of basic seismic design principles, both force-based and displacement-based conventional design.

Seismic isolation bearings were discussed in Chapter 8, Section 8.3, and seismic displacement capacity calculations were discussed in Chapter 9, Section 9.7. Those discussions will not be repeated here, but the reader will find occasional reference to those sections informative while going through the material in the current chapter. Additionally, Chapter 11 introduces additional material on seismic isolation of bridges.

Seismic design of bridges may be accomplished using either (a) force-based design in accordance with the LRFD BDS (AASHTO, 2020), or (b) displacement-based design in accordance with the LRFD GS (AASHTO, 2011).

Displacement-based design is generally accepted as the more appropriate engineering approach, given that seismic design involves loading beyond the elastic range.

Section 2.8 of this text presented a discussion on the development of seismic loads in terms of a design response spectrum and the assignment of a Seismic Zone (LRFD BDS) or a Seismic Design Category (LRFD GS). Each design method will now be discussed.

This chapter deals with conventional design, based on either (a) plastic hinge development in substructure columns or (b) ductile cross-frames with essentially elastic substructures. For discussion of a third seismic design strategy, seismic isolation, refer to Chapter 11 of this text.

While typical design procedures incorporate linear response spectrum analysis techniques, design by nonlinear response history analysis is also permitted. Response history analysis requires the development of not only a design response spectrum, but also a suite of appropriately selected and modified ground motion record pairs. Hence, this Chapter also includes a detailed discussion of ground motion selection and modification for inelastic response history analysis.

Design ground motion in the current AASHTO LRFD BDS is based on geometric mean, uniform hazard spectra. The likelihood is high that the near future will find changes in design earthquake ground motion for bridges. For example, ASCE 7-16 for buildings now adopts risk-targeted (rather than uniform hazard), maximum direction (RotD100, rather than geometric mean) ground motion as the design basis. Additionally, design response spectra have typically been generated using three control points (A_S, S_{DS}, and S_{D1}) to define the entire design response spectrum. These data are available from the USGS. But the future may find bridge engineers using a 22-point (rather than a three-point) design response spectrum, with site condition effects explicitly incorporated into the mapped values of spectral acceleration. Such a tool is currently available, in beta version, at the USGS (https://earthquake.usgs.gov/ws/designmaps/nehrp-2020.html).

It is critical that bridge engineers maintain design procedures aligning with the most up-to-date seismic loading definitions.

10.1 FORCE-BASED SEISMIC DESIGN BY THE LRFD BDS

In force-based seismic design by the LRFD BDS, a force-reduction factor is first established. This requires that the owner specifies an importance category for each particular bridge project. These categories include (a) critical, (b) essential, and (c) other. These categories are likely to change in the near future, but, as of May 2021, are as specified herein. For design by response spectrum analysis, force-reduction (R) factors are summarized in Table 10.1. For design by inelastic response history analysis, $R = 1.0$. With inelastic response history analysis, inelasticity is directly and explicitly modeled.

For the connection of ductile columns or pile bent piles to the cap and foundation, it is permitted to use capacity protection procedures in which overstrength capacity of ductile elements is determined to establish the maximum load which can physically be transmitted to the connection or connected element. The capacity protection method is preferred and may be used in lieu of the R-factors for connections.

Once determined by an acceptable analysis technique, elastic seismic forces are reduced by the factor, R, and combined for orthogonal, horizontal effects as follows to account for bi-directional ground motions.

- 100 percent of the design force in direction 1 combined with 30 percent of the design force in direction 2
- 100 percent of the design force in direction 2 combined with 30 percent of the design force in direction 1

TABLE 10.1
Force-Reduction (R) Factor for Force-based Seismic Design

Substructure	Critical	Essential	Other
Wall-type piers	1.5	1.5	2.0
Concrete pile bents, vertical piles only	1.5	2.0	3.0
Concrete pile bents, batter piles	1.5	1.5	2.0
Single-column bents	1.5	2.0	3.0
Steel and CFST pile bents, vertical piles only	1.5	3.5	5.0
Steel and CFST pile bents, batter piles	1.5	2.0	3.0
Multi-column bents	1.5	3.5	5.0
Connection, superstructure to abutment	0.8	0.8	0.8
Within-span expansion joints	0.8	0.8	0.8
Connection, columns, or pile bents to cap	1.0	1.0	1.0
Connection, columns, or piles to foundation	1.0	1.0	1.0

When non-ductile, capacity-protected elements are designed for the expected, over-strength capacity of the ductile elements, such combination is unnecessary. The overstrength plastic shear must simply be resisted in any direction.

For bridges in Seismic Zone 1 with A_S less than 0.05, the design force in any direction is to be taken no less than 15 percent of the total vertical reaction due to combined (a) permanent loads and (b) live load assumed to act concurrently with the earthquake. For other bridges in Seismic Zone 1, the design force in any direction is to be taken as not less than 25 percent of the total vertical reaction.

For Seismic Zone 2, capacity-protected, non-ductile elements are designed for a force-reduction factor, R, equal to one-half of the value from Table 10.1, but no less than 1.0. Once again, an alternative is to design for expected, overstrength capacity of the ductile elements.

For Seismic Zones 3 and 4, hinging, ductile elements (usually the columns in a multi-post bent, or the piles in a pile bent) are first designed using the reduced seismic forces. Capacity-protected, non-ductile elements are then designed for either (a) unreduced elastic seismic forces ($R = 1.0$) or (b) the inelastic hinging forces. Inelastic hinging forces are to be determined using resistance factors, $\phi = 1.3$ for concrete or $\phi = 1.25$ for structural steel, to account for overstrength. This is an estimate of the plastic shear for a bent or pier. The plastic shear is the largest load which may be transmitted from a ductile element to the connected elements. Capacity design for plastic hinging forces is to be greatly preferred over design based on unreduced seismic forces, in the author's opinion.

Section 5.11 of the LRFD BDS contains specific seismic requirements for concrete elements. For details not covered here, the reader is referred to that section.

Requirements outlined here are generally those applicable to Seismic Zones 3 and 4, as those requirements are more fully developed and appropriate for the seismic design of bridges, in the author's opinion.

For circular columns and piles, Equation 10.1 provides the required volumetric spiral or seismic hoop reinforcement for plastic hinge regions. The core diameter, d_c, is measured to the outside diameter of the spiral or hoop. The pitch, s, is measured vertically to the center of the spiral or hoops. A_{sp} is the cross-sectional area of the spiral or hoop bar. The specified minimum yield strength, f_{yh}, is that for the spiral or hoop bar, but not necessarily the same as that for longitudinal bars. The specified concrete strength, f'_c, is used for the calculations, not the expected concrete strength, f'_{ce}. Equation 10.2 from Section 5.6.4.6 of the LRFD BDS for non-seismic confinement requirements, should always be checked as well. The core area, A_c, is based on the diameter to the outside of the spiral or hoop as well.

Equations 10.3 and 10.4 provide the required confinement reinforcement, A_{sh}, within a spacing equal to s, for rectangular piles and columns. The dimension, h_c, is measured to the outside of the transverse bars in the direction of loading. The spacing, s, is to exceed neither (a) 4 inches, nor (b) one-quarter of the least cross-sectional dimension.

$$\rho_s = \frac{4A_{sp}}{d_c s} \geq 0.12 \frac{f'_c}{f_{yh}} \tag{10.1}$$

$$\rho_s = \frac{4A_{sp}}{d_c s} \geq 0.45 \left(\frac{A_g}{A_c} - 1 \right) \frac{f'_c}{f_{yh}} \qquad (10.2)$$

$$A_{sh} \geq 0.12 s h_c \frac{f'_c}{f_{yh}} \qquad (10.3)$$

$$A_{sh} \geq 0.30 s h_c \left(\frac{A_g}{A_c} - 1 \right) \frac{f'_c}{f_{yh}} \qquad (10.4)$$

For either round or rectangular columns and piles, the yield strength of the transverse reinforcement, f_{yh}, is the minimum specified value for the reinforcement used, but is not to exceed 75 ksi.

10.2   DISPLACEMENT-BASED SEISMIC DESIGN BY THE LRFD GS

Displacement-based seismic design involves (a) identification of ductile elements to be designed and detailed for inelastic behavior, (b) ensurance that the displacement demand on each ductile element is less than its displacement capacity, and (c) capacity-protection of elements intended to remain essentially elastic during strong ground shaking.

Displacement ductility is defined as the maximum displacement experienced by an element divided by the yield displacement of the element. For pile bent design by the displacement-based provisions of the LRFD GS, the displacement ductility demand, μ_D, is not to exceed 4. For single-column bents, μ_D is limited to 5, and for multi-columns bents with above-ground hinging, μ_D is limited to 6.

Displacement capacity has been discussed in Section 9.7 of this text. Displacement demand may be determined by (a) equivalent static analysis with displacement amplification for short-period structures, (b) multi-mode elastic response spectrum analysis with displacement amplification for short-period structures, or (b) inelastic response history analysis. Orthogonal component combination for methods (a) and (b) are based on the 100-30-30 rule, while orthogonal component interaction is inherently accounted for in inelastic response history analysis, when three-dimensional modeling is adopted. Refer to the LRFD GS, Section 4.2, for guidance on permissible analysis methods for various bridge complexity levels.

For equivalent static and elastic response spectrum methods, short-period displacement amplification is required for natural modes having periods less than or equal to $1.25T_S$. See Section 2.8 on earthquake loading for the definition of T_S. The required amplification, R_d, of displacement demand is given by Equation 10.5. Design response spectra are typically based on damping equal to 5 percent of critical. In rare cases for which larger damping is appropriate, Equation 10.6 is provided in the LRFD GS for modification of the design response spectrum across all periods by the factor, R_D.

$$R_d = \left(1 - \frac{1}{\mu_D} \right) \left(\frac{1.25T_S}{T} \right) + \frac{1}{\mu_D} \qquad (10.5)$$

$$R_D = \left(\frac{0.05}{\xi}\right)^{0.40} \tag{10.6}$$

For Equation 10.5, μ_D is equal to 2 for Seismic Design Category B, 3 for Seismic Design Category C, and determined by analysis for Seismic Design Category D. For Seismic Design Category D, μ_D may conservatively be taken to be equal to 6 in Equation 10.5.

Second-order $P\Delta$ effects may be ignored in structural analysis for earthquake loading whenever Equation 10.7 is satisfied for concrete substructures, or Equation 10.8 for steel substructures. The unfactored dead load is P_{dl}. M_p is the idealized plastic moment determined using expected, rather than specified minimum, material properties. M_n is the nominal moment capacity based on nominal properties. Presumably, Δ_r is the first-order displacement between points of contraflexure and plastic hinging (approximated by one-half of the total displacement for a rigid frame with hinges at top and bottom of columns).

$$P_{dl}\Delta_r \le 0.25M_p \tag{10.7}$$

$$P_{dl}\Delta_r \le 0.25M_n \tag{10.8}$$

Equation 10.9 for analytical plastic hinge length, L_P, is appropriate for reinforced concrete columns framing into oversized shafts, footings, cased shafts, and bent caps. In the equation, L is the distance from the point of maximum moment to the point of contraflexure, not necessarily equal to the clear column height. The diameter, d_{bl}, is that for the longitudinal bars in the column. For reinforced and prestressed concrete piling, and for cast-in-drilled-hole shafts, Equation 10.10a provides the analytical plastic hinge length. H' is the distance from ground surface to the point of above-ground contraflexure. D^* is the diameter for a circular member, or otherwise the cross-sectional dimension in the direction of loading. For concrete-filled steel tube pipe piles of diameter D, Equation 10.10b is specified in the LRFD GS.

$$L_P = 0.08L + 0.15f_{ye}d_{bl} \ge 0.30f_{ye}d_{bl} \tag{10.9}$$

$$L_P = 0.1H' + D^* \le 1.5D^* \tag{10.10a}$$

$$L_P = 0.1H' + 1.25D \le 2.0D \tag{10.10b}$$

The plastic hinge region for reinforced concrete columns and piles, within which enhanced lateral confinement reinforcement is to be provided, is taken as the greater of:

- 1.5 times the dimension in the direction of loading
- the region where moment demand exceeds 75 percent of the plastic moment
- the analytical plastic hinge length, L_P

Minimum support lengths at expansion joints are determined from Equation 10.11 for Seismic Design Category A. L is the length, in feet, of the bridge to the adjacent expansion joint or the end of the bridge. N is the minimum required support length, in inches. S is the skew of the deck to substructure alignment, in degrees. H is the average height, in feet, of columns supporting the bridge deck at piers from abutment to abutment.

$$N = \left(8 + 0.02L + 0.08H\right)\left(1 + 0.0001255S^2\right) \tag{10.11}$$

For Seismic Design Categories B and C, the length given by Equation 10.11 is to be increased by a factor of 1.5.

For Seismic Design Category D, the required support length is a function of the displacement demand from the structural analysis (Δ_{eq}), and is given by Equation 10.12. For single-span bridges in Seismic Design Category D, the support length is taken to be not less than 1.5 times the value from Equation 10.11. The skew effect multiplier in Equation 10.12 is different from that in Equation 10.11. It is not clear if this is intentional in the LRFD GS. Nonetheless, the equations presented here are taken from the LRFD GS.

$$N = \left(4 + 1.65\Delta_{eq}\right)\left(1 + 0.00025S^2\right) \geq 24 \tag{10.12}$$

For structural analysis, abutment longitudinal stiffness (K_{eff}) and strength (P_p) may be ignored, with intermediate piers designed to resist all seismic effects, or may be included (when permitted by the owner) as given in Equations 10.13 and 10.14.

$$P_p = H_w W_w p_p \tag{10.13}$$

$$K_{eff1} = \frac{P_p}{F_w H_w} \tag{10.14}$$

- K_{eff} = abutment stiffness, kips/ft
- P_p = passive capacity, kips
- p_p = presumptive passive pressure, ksf
- H_w = backwall height, ft
- W_w = backwall width, ft
- F_w = factor ranging from 0.01 (dense sand) to 0.05 (compacted clay)

Unless special attention, beyond that normally provided for backfill, is introduced into the plans and specifications, 70 percent of the presumptive passive pressure, p_p, should be used in the analysis. Refer to the LRFD GS, Section 5.2.3.1, for additional information. Presumptive pressure, if used in lieu of detailed analysis procedures, requires that the backfill be compacted to a dry density greater than 95 percent of the maximum, and is specified in the LRFD GS as follows:

- for cohesionless, non-plastic backfill, p_p may be taken equal to $2H_w/3$ ksf per foot of wall length.

- provided that the estimated undrained shear strength is greater than 4 ksf, p_p for cohesive backfill may be taken to be equal to 5 ksf.

Equation 10.14 provides an initial stiffness estimate, which may be used in a response spectrum or response history analysis in cases for which abutment stiffness and strength is to be used to resist seismic forces. From the initial structural analysis, the longitudinal forces at the abutments will be available. Should this force exceed P_p, the initial spring stiffness should be softened and the analysis re-run. The iterative procedure should be performed until the assumed stiffness is consistent with the computed stiffness.

10.3 CAPACITY DESIGN PRINCIPLES

Displacement-based and force-based design both require attention to capacity-design principles. Ductile elements are detailed to ensure inelastic deformation capacities in excess of the estimated required deformation during strong ground shaking. Other elements are designed for a higher force level, thus ensuring their essentially elastic behavior during the event. For bridge substructures, when column plastic hinging is the design strategy, overstrength plastic shear calculations are made and all other elements are designed for the expected, overstrength plastic shear. Some of these capacity design provisions from the AASHTO LRFD GS are summarized here. Readers are referred to the LRFD GS for a complete treatment of these principles and requirements.

Plastic hinging forces are to be computed using expected (not minimum) material strengths with an additional over-strength factor. The over-strength factor, λ_{mo}, is determined as follows:

- $\lambda_{mo} = 1.4$, reinforced concrete hinging columns with A 615 reinforcing
- $\lambda_{mo} = 1.2$, reinforced concrete with A 706 reinforcing
- $\lambda_{mo} = 1.2$, structural steel hinging columns

Expected material strengths are $f'_{ce} = 1.3 f'_c$ for concrete and 68 ksi for the yield stress, f_{ye}, of steel reinforcing bars, whether A615 Grade 60 or A706 Grade 60. Expected tensile strength, f_{ue}, is 95 ksi for both bar specifications. For hinging columns in Seismic Design Category D, A706 reinforcement is required in lieu of A615. Strain-hardening effects are to be incorporated into the reinforcing steel material model.

Strain limits in reinforcement are summarized below, with recommended ultimate tensile strains on the order of two-thirds to three-fourths of actual minimum values for added safety. The onset of strain hardening is defined by ε_{sh}, and the ultimate tensile strain by ε^R_{su}.

For the onset of strain hardening:

- $\varepsilon_{sh} = 0.0150$, A 615 and A 706 bars, #3–#8
- $\varepsilon_{sh} = 0.0125$, A 615 and A 706 bars, #9
- $\varepsilon_{sh} = 0.0115$, A 615 and A 706 bars, #10–#11

- $\varepsilon_{sh} = 0.0075$, A 615 and A 706 bars, #14
- $\varepsilon_{sh} = 0.0050$, A 615 and A 706 bars, #18

For bar fracture:

- $\varepsilon^R_{su} = 0.090$, A 706 bars, #4–#10
- $\varepsilon^R_{su} = 0.060$, A 615 bars, #4–#10
- $\varepsilon^R_{su} = 0.060$, A 706 bars, #11–#18
- $\varepsilon^R_{su} = 0.040$, A 615 bars, #11–#18

Limiting compressive strains in the concrete core are typically determined using the Mander model for confined concrete (see Section 9.7 of this text) and depend on the amount of transverse reinforcement provided in the form of hoops, spirals, or rectangular ties. Core compressive strains at in-ground plastic hinges, when permitted, should be limited to 0.02.

Determination of the plastic shear for a multi-column bent or pier is an iterative procedure. Problem 10.2 provides a detailed example, including confinement calculations, displacement demand calculations, displacement capacity calculations, and plastic shear calculations. Section analysis software is a requirement for such analyses.

Capacity design provisions for footings may be found in the AASHTO LRFD GS, Sections 6.4.5, 6.4.6, and 6.4.7. For a footing of length L in the direction of loading, column dimension D_c in the direction of loading, and depth H_f, Equation 10.15 must be satisfied in order to consider the footing or pile cap rigid. Otherwise, non-standard analyses are required to determine pressure distribution for a footing, or pile loads for a pile cap.

Equation 10.16 is the overturning analysis requirement for a spread footing. M_{po} is the overstrength plastic moment of the column and V_{po} is the overstrength plastic shear. The axial force associated with the overstrength plastic hinging forces is P_u and the resistance factor, ϕ, for overturning, is 1.0. The nominal bearing capacity of the supporting material is q_n. The footing width, perpendicular to the direction of loading, is B. The footing width effective in resisting flexure and shear, b_{eff}, is given by Equation 10.18. The column dimension perpendicular to the direction of loading is B_c.

$$\frac{(L - D_c)}{2H_f} \leq 2.5 \tag{10.15}$$

$$M_{po} + V_{po}H_f \leq \phi P_u\left(\frac{L - a}{2}\right) \tag{10.16}$$

$$a = \frac{P_u}{q_n B} \tag{10.17}$$

$$b_{eff} = B_c + 2H_f \leq B \tag{10.18}$$

For additional footing and pile cap requirements not covered here, refer to the LRFD GS, Sections 6.3 and 6.4.

Capacity design provisions for non-integral bent caps extensive and are found in Sections 8.12 and 8.12 of the LRFD GS. The reader is referred to these provisions for detailed design requirements.

10.4   GROUND MOTION SELECTION AND MODIFICATION FOR RESPONSE HISTORY ANALYSIS

For structural design by response history analysis, it becomes necessary to select a suite of ground motion records. Typical suite sizes range from as few as three record pairs to as many as 11 or more. In fact, given the emphasis on performance-based design likely to occur in the near future, as many as 30 or 40 record pairs (or more) could be a necessary minimum in research when estimates of response variability to ground shaking are needed.

Ground motion selection requires careful attention to several factors. An excellent reference for the process is found in NIST-GCR-11-917-15 (NEHRP Consultants Joint Venture, 2011). Engineers and researchers working on ground motion selection would be well served in consulting this freely available digital document.

Some of the parameters involved in selecting candidate ground motions for structural analysis include, in descending order of importance (in the author's opinion), at least for non-subduction earthquakes:

1. match to spectral shape
2. magnitude
3. recording station site characterization
4. distance
5. fault type

Given that match to spectral shape is a critical factor in whether or not a particular ground motion should be considered for a given site, some means of measuring match to spectral shape is necessary. Two proposed measures of match to spectral shape are mean-square-error (MSE) and D_{RMS}. MSE and D_{RMS} are given in Equations 10.19 (Pacific Earthquake Engineering Research Center, 2010) and 10.20. Notice that MSE has a scale factor, f. Equation 10.21 provides an expression for computing the scale factor, f, which minimizes MSE over a specified range of periods. This is the scale factor used in amplitude scaling of ground motion records discussed in the section on ground motion modification. So MSE can be computed pre-scaling by inserting $f = 1$ in the equation for MSE, and post-scaling by inserting the computed scale factor for f in the equation for MSE. The post-scaled MSE is the appropriate value for assessing candidate records. The weights, $w(T_i)$, at each period are typically taken to be equal to 1 for all periods in the range of interest.

$$MSE = \frac{\sum w(T_i) \cdot \left\{ \ln\left[SA_{TARGET}(T_i) \right] - \ln\left[f \cdot SA_{GM}(T_i) \right] \right\}^2}{\sum w(T_i)} \tag{10.19}$$

$$D_{RMS} = \frac{1}{N} \sqrt{\sum_{i=1}^{N} \left(\frac{(SA_{GM})_{Ti}}{PGA_{GM}} - \frac{(SA_{TAR})_{Ti}}{PGA_{TAR}} \right)^2} \qquad (10.20)$$

$$\ln f = \frac{\sum \left[w(T_i) \cdot \ln \dfrac{SA_{TARGET}(T_i)}{SA_{GM}(T_i)} \right]}{\sum w(T_i)} \qquad (10.21)$$

A knowledge of characteristic magnitude and distance combinations may be obtained for a given site by disaggregating the seismic hazard, available at the USGS online Unified Hazard Tool (https://earthquake.usgs.gov/hazards/interactive/).

Once a set of candidate records (perhaps 100 or more) has been established, it is necessary to have a means of (a) selecting the most appropriate records and (b) modifying the records. While many candidate records from a single event may be included, it is customary (and sometimes required) that no more than three or four records from any single event be included in the final suite.

There are at least three methods for modifying ground motion records for use in structural analysis:

1. amplitude scaling
2. spectral matching in the time domain
3. spectral matching in the frequency domain

Amplitude scaling is typically the preferred method for ground motion modification. The accelerations at each time step in the record are all multiplied by an appropriate factor, such as that computed from Equation 10.17. The time scale is not adjusted in any fashion. Frequency content and pulse-type character of the ground motion are retained.

The PEER Ground Motion Database (Pacific Earthquake Engineering Research Center, 2014) includes an excellent online tool for ground motion selection with subsequent amplitude scaling. Software for ground motion scaling includes *SigmaSpectra* (Albert Kottke), which is freely available, and *SeismoSelect*, for which free educational licenses are available.

SigmaSpectra (https://github.com/arkottke/sigmaspectra) offers the advantage of permitting the user to specify a target log-based standard deviation in addition to a target pseudo-spectral acceleration (PSA) spectrum. ASCE 7-16 (Section 21.2.1.2) suggests a log-based standard deviation equal to 0.60 across all periods of interest. This presumably preserves record-to-record variability with a small subset of all available ground motions, and could be used to design to achieve responses greater than the median (by some number, say 1, of standard deviations). Ground motion models may also be used to establish target, log-based variability.

SeismoSelect (www.seismosoft.com) provides another alternative for ground motion scaling. Included in the software are various options for target basis (RotD100,

GeoMean, etc.) as well as for ground motion database sources (PEER, ESMD, etc.). *SeismoSelect* also has features which enable the user to generate code-based target spectra for many different specifications.

For spectral matching in the time domain, *SeismoMatch* (SeismoSoft, 2020b) adds wavelets to an accelerogram to produce a new accelerogram to which the *PSA* response spectrum matches, as closely as possible, the target spectrum. Modification is performed directly on the accelerogram in the time domain.

SeismoArtif (SeismoSoft, 2020a) has multiple capabilities, among which is spectral matching in the frequency domain. The Fourier spectrum for a record is first computed and compared to a Fourier spectrum generated from the target *PSA* spectrum. Adjustment to the ground motion Fourier spectrum is made to more closely match that of the target Fourier spectrum. The modified Fourier spectrum is then converted back to a new accelerogram.

Care must be taken in spectral matching, whether time-domain-based or frequency-domain-based, to ensure that realistic (if somewhat subjective) ground motions are produced in the matching process. Any pulse-type character for near-fault effects required by a governing code or specification may be significantly altered when spectral matching is employed, but is preserved when amplitude scaling is used.

In addition to modification of real ground motion records, artificial and synthetic records may, at times, be useful. Real records, modified by amplitude scaling, are the preferred choice, in the author's opinion. Nonetheless, a brief discussion of artificial and synthetic record options is appropriate.

Physics-based models exist for generating synthetic ground motions. *SeismoArtif* (SeismoSoft, 2020a) has this capability, in addition to those previously mentioned. Near-fault and far-field options are available. Site conditions and tectonic environment (intraplate *vs.* plate boundary) are specified by the user. Both synthetic accelerogram generation/adjustment and artificial accelerogram generation/adjustment are available. In the context adopted here, synthetic accelerograms are those based on physics-based models. Artificial accelerograms are those generated as a random signal and bounded with a specified enveloping function.

A simple model for synthetic record generation is available through the State University of New York (SUNY) program *RSCTH*. Required input is minimal, including moment magnitude, epicentral distance, target spectrum parameters, and tectonic regime (Eastern US or Western US). The program is DOS based.

For the target response spectrum, it is necessary to assign a period range of interest. Design specifications vary in defining this period range. While appropriate for buildings, ASCE 7-16 may be the preferable standard available in defining an appropriate period range of interest. ASCE 7-16 defines the period range of interest as follows.

- The period range of interest should have an upper bound greater than or equal to twice the larger of the two fundamental periods in the two orthogonal, horizontal directions. In the opinion of the author, this should be extended to 2.5 times the larger period, given that the effective period of an elastic-perfectly-plastic structure having a displacement ductility demand of 6 (the largest value permitted for conventional design in the AASHTO LRFD GS) is equal to the elastic period multiplied by the square root of 6 (i.e., 2.45).

- The period range of interest should have a lower bound equal to the smaller of (a) the period required to achieve at least 90% mass participation in each horizontal, orthogonal direction and (b) 20% of the smallest fundamental period in the two horizontal, orthogonal directions.

For ground motion selection requirements, AASHTO is relatively silent. However, the FHWA retrofit manual for bridges (Buckle, et al., 2006) does contain a few requirements. These are summarized below.

- Either three ground motions or seven ground motions may be used in the analysis. If three ground motions are used, the maximum response of the three should be used for design. If seven are used, the average response may be taken as the design value.
- The suite mean spectrum should not fall below the target by more than 15% over the period range of interest, and the average ratio of suite mean to target over the entire period range of interest should be at least 1.0.
- Amplitude scaled or spectrally matched records are permissible.
- For near-fault sites, ground motions are to be rotated to fault-normal (FN) and fault-parallel (FP) orientations for application in structural analysis.

Although not strictly required by current AASHTO provisions, it seems wise to adopt ground motion suites with no fewer than eleven record pairs for bridge design, given that the ASCE provisions for suite size are more developed than those in AASHTO.

Figure 10.1 depicts an example of an appropriately generated ground motion suite. The plot clearly illustrates that more stringent criteria than those found in the FHWA retrofit manual or in AASHTO are generally satisfied by the suite of record pairs. Such criteria may include each of the following, depending on the design specification adopted.

- For amplitude scaling, record pairs should be scaled such that the suite mean *PSA* spectrum generally matches or exceeds the target spectrum over the period range of interest. This may be quantitatively evaluated by averaging the suite-mean *PSA* to target *PSA* ratio over the period range of interest.
- For amplitude scaling, record pairs should be scaled such that the suite mean *PSA* spectrum does not fall below 90% of the target spectrum at any period within the range of interest.
- For amplitude scaling, record pairs should be scaled such that the suite mean *PSA* spectrum does not exceed 130% of the target spectrum at any period within the range of interest.

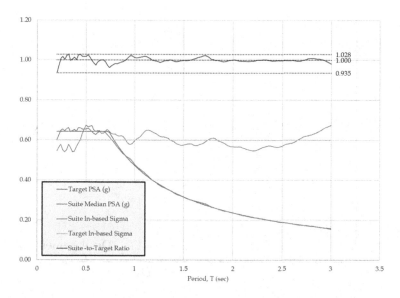

FIGURE 10.1   Example ground motion suite spectra.

- The target, log-based standard deviation on *PSA* should equal 0.60 over the period range of interest. Ground motion models may also be useful in establishing target, log-based standard deviation.

In Figure 10.1, the period range of interest is [0.0–3.0] seconds. The minimum PSA_{Suite}/PSA_{Target} is 0.935 (greater than 0.9 required). The average PSA_{Suite}/PSA_{Target} is 1.000 (greater than or equal to 1.0, and preferable very near 1.0). The maximum PSA_{Suite}/PSA_{Target} is 1.028 (less than 1.3 as required). Over the period range of interest, the average σ_{lnPSA} is 0.599 (very near the target 0.60).

10.5   SUBSTITUTE-STRUCTURE METHOD (SSM) ANALYSIS

The substitute-structure method (SSM) for seismic analysis is often a valuable tool in estimating seismic displacement demands. Although the final design for critical structures is likely to require nonlinear response history analysis for final design details, SSM analysis has been shown to provide reliable seismic displacement estimates (Huff and Pezeshk, 2016). The method, originally proposed by Gulkan and Sozen (1974), is the basis for the preliminary design of isolation devices in the AASHTO GS ISO (AASHTO, 2014).

SSM analysis is a response spectrum-based procedure and requires no ground motion selection. Only the design response spectrum is required for loading definition. To complete the analysis, secant stiffness (K_{EFF}) at maximum displacement is used to compute an effective period (T_{EFF}, Equation 10.22). Figure 10.2 depicts the essential parameters in an assumed bi-linear force displacement relationship.

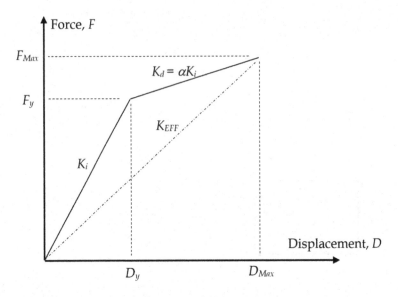

FIGURE 10.2   Bi-linear force-displacement in the SSM analysis.

Models for (a) added damping and (b) response modification due to added damping are required as well. Displacement ductility, μ_D, is defined by Equation 10.23.

Various damping (ξ_{EFF}) models include those proposed by Priestley et al. (2007). Equation 10.24 has been proposed as valid for reinforced concrete bridge substructures under inelastic displacement demands. Equation 10.25 is the theoretical elastic plus hysteretic damping for any bilinear oscillator. For steel framed buildings, Equation 10.26 was proposed by Priestley et al. (2007).

Models for response modification (R_ξ) as a result of added damping include Equation 10.27 for non-pulse type ground motion and Equation 10.28 for pulse-type ground motion, both from Euro Code 8. The equivalent of Equation 10.29 is found in the AASHTO GS ISO (AASHTO, 2014) for seismic isolation devices. A model which depends on both effective damping and significant duration (D_{5-95}) has been developed as well (Stafford et al., 2008), given by Equation 10.30.

The structure is treated as a single-degree-of-freedom (SDOF) oscillator in the SSM. From the design response spectrum, PSA_{TEFF} is the pseudo-spectral acceleration at the effective period of the SDOF. The displacement demand is given by Equation 10.31.

Careful observation of the equations will make it clear that the displacement demand, D_{Max}, depends on the effective damping, ξ_{EFF}, which depends on D_{Max}. Hence, the iterative nature of the SSM becomes evident. Hence, depending on the type of problem, it may be necessary to assume an initial value for either μ_D, D_{Max}, or ξ_{EFF}, performing subsequent iterations until the calculated results agree with the assumption.

$$T_{EFF} = 2\pi \sqrt{\frac{W}{gK_{EFF}}} = T_i \sqrt{\frac{\mu_D}{1+\alpha\mu_D - \alpha}} \tag{10.22}$$

$$\mu_D = \frac{D_{Max}}{D_y} \tag{10.23}$$

$$\xi_{EFF} = 0.05 + 0.444 \frac{\mu_D - 1}{\pi\mu_D} \tag{10.24}$$

$$\xi_{EFF} = \xi_{EL} + \frac{2(\mu_D - 1)(1-\alpha)}{\pi\mu_D(1+\alpha\mu_D - \alpha)} \tag{10.25}$$

$$\xi_{EFF} = 0.05 + 0.577 \frac{\mu_D - 1}{\pi\mu_D} \tag{10.26}$$

$$R_\xi = \left(\frac{0.10}{0.05 + \xi_{EFF}}\right)^{0.50} \tag{10.27}$$

$$R_\xi = \left(\frac{0.07}{0.02 + \xi_{EFF}}\right)^{0.25} \tag{10.28}$$

$$R_\xi = \left(\frac{0.05}{\xi_{EFF}}\right)^{0.30} \geq 0.588 \tag{10.29}$$

$$R_\xi = 1 - \frac{-0.631 + 0.421\ln(\xi_{EFF}) - 0.015\ln(\xi_{EFF})^2}{1+\exp\left\{\frac{-[\ln(D_{5-95}) - 2.047]}{0.930}\right\}} \tag{10.30}$$

$$D_{Max} = (PSA_{TEFF} \cdot g)\left(\frac{T_{EFF}}{2\pi}\right)^2 \cdot R_\xi \tag{10.31}$$

In approximate analyses using SSM techniques, it may at times be desirable to estimate effects due to plan torsional irregularities. The method outlined in ASCE 7-16, Section 17.5.3.3, can be useful for such estimates. Equation 10.32 provides an estimate for displacement amplification due to plan torsional offset of the center of mass, often assumed to be 5% or more of the plan dimension. Eccentricity, e, is defined to be the distance between the centers of mass and rigidity. The plan dimensions are b and d. The distances from the center of mass to each isolator unit along two perpendicular axes are x and y. The distance from the center or rigidity to an isolator in question is z.

$$D'_{Max} = D_{Max}\left[1 + \frac{z}{P_T^2} \cdot \frac{12e}{b^2 + d^2}\right] \geq 1.15 D_{Max} \tag{10.32}$$

$$P_T = \frac{1}{r}\sqrt{\frac{\Sigma\left(x^2 + y^2\right)}{N}} \tag{10.33}$$

$$r = \sqrt{\frac{b^2 + d^2}{12}} \tag{10.34}$$

10.6 SHEAR RESISTANCE AT THE EXTREME EVENT LIMIT STATE

For reinforced concrete columns subject to large seismic shear, conventional resistance models may prove unreliable. Multiple methods exist for the evaluation of shear resistance at the Extreme Event limit state for columns which form plastic hinges during strong ground shaking.

The AASHTO LRFD GS adopts Equation 10.35 for design shear resistance within the plastic hinge region and recommends a resistance factor $\phi = 0.90$ rather than the value of 1.0 used for most Extreme Event limit states. The same equation may be used for ductile columns outside the plastic hinge region by setting α' equal to 3. The concrete shear stress, v_c, is calculated from Equation 10.37 if the factored axial load, P_u, is compressive. Otherwise, $v_c = 0$. Equations 10.38, 10.39, and 10.40 are applicable for concrete resistance of round columns. Equations 10.41, 10.42, and 10.43 are applicable for concrete resistance of rectangular columns. For the shear resistance provided by reinforcement, Equation 10.44 is for round columns and Equation 10.45 is for rectangular columns.

$$\phi V_n = 0.90\left(V_c + V_s\right) \tag{10.35}$$

$$V_c = v_c\left(0.80 A_g\right) \tag{10.36}$$

$$v_c = 0.032\alpha'\left(1 + \frac{P_u}{2A_g}\right)\sqrt{f_c'} \leq \min\begin{cases} 0.11\sqrt{f_c'} \\ 0.047\alpha'\sqrt{f_c'} \end{cases} \tag{10.37}$$

$$\alpha' = \frac{f_s}{0.15} + 3.67 - \mu_D \tag{10.38}$$

$$f_s = \rho_s f_{yh} \leq 0.35 \tag{10.39}$$

$$\rho_s = \frac{4A_{sp}}{sD'} \tag{10.40}$$

$$\alpha' = \frac{f_w}{0.15} + 3.67 - \mu_D \qquad (10.41)$$

$$f_w = 2\rho_w f_{yh} \le 0.35 \qquad (10.42)$$

$$\rho_s = \frac{A_v}{sb} \qquad (10.43)$$

$$V_s = \frac{\pi}{2}\left(\frac{nA_{sp}f_{yh}D'}{s}\right) \le 0.25\sqrt{f_c'}\left(0.80A_g\right) \qquad (10.44)$$

$$V_s = \frac{A_v f_{yh} d}{s} \le 0.25\sqrt{f_c'}\left(0.80A_g\right) \qquad (10.45)$$

- A_g = gross cross-sectional area, in^2
- P_u = factored compressive force, kips
- A_{sp} = area of spiral or hoop bar, in^2
- s = pitch of spiral, hoops, or ties, inches
- D' = core diameter measured to center of spiral or hoop, inches
- A_v = total area of shear reinforcement in the direction of loading, in^2
- b = width of rectangular member perpendicular to direction of loading, inches
- f_{yh} = nominal yield stress of transverse reinforcement, ksi
- f_c' = nominal compressive strength of concrete, ksi
- μ_D = local displacement ductility demand for the member in question
- α' = adjustment factor
- n = number of individual interlocking cores with spiral or hoop bars
- d = effective depth in the direction of loading, inches

The effective depth, d, for rectangular members is measured from the extreme compression face to the centroid of the tensile reinforcement.

For Seismic Design Category B, shear resistance within the plastic hinge zone is to be determined using $\mu_D = 2$. For Seismic Design Category C, shear resistance within the plastic hinge zone is to be determined using $\mu_D = 3$. For Seismic Design Category D, shear resistance within the plastic hinge zone is to be determined using μ_D as determined from analysis.

Minimum transverse reinforcement must be provided in accordance with the following:

- for Seismic Design Category B, $\rho_s \ge 0.003$ and $\rho_w \ge 0.002$
- for Seismic Design Category C and D, $\rho_s \ge 0.005$ and $\rho_w \ge 0.004$

For members with interlocking cores and spiral or hoop reinforcement, the shear reinforcement resistance is taken as the sum of the individual contributions from each interlocking spiral or hoop.

For details which are not covered here, the reader is referred to the LRFD GS, Section 8, and to the literature (Priestley et al., 2007) for additional shear models.

10.7 SOLVED PROBLEMS

Problem 10.1

The pile bent shown in Figure P10.1 is used as an intermediate pier in a short-span continuous bridge. The CFST piles are 20-inch diameter with 5/8-inch wall thickness and are fabricated in accordance with ASTM A53 Grade B, $F_y = 35$ ksi. The concrete core has a 28-day compressive strength, $f'_c = 3$ ksi. For the condition shown, each of the five girder-bearing locations has a reaction of 117 kips.

The cap is 4 ft by 4 ft in cross section. The lateral load shown represents an incremental load to be applied at the superstructure center of gravity in a pushover analysis.

Determine the plastic shear and the displacement capacity from push-over analysis. *SeismoStruct* will be used for the analysis.

Problem 10.2

A bridge pier is to be constructed at a seismic zone 4 (Seismic Design Category D) site in the NMSZ. The geometry and data for the bridge piers

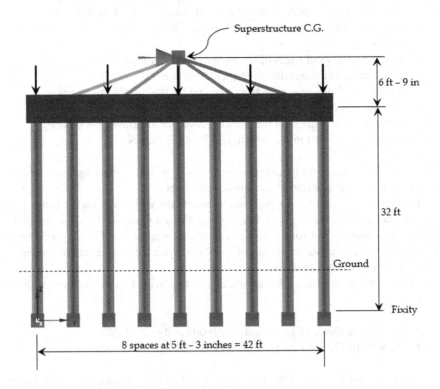

FIGURE P10.1 Problem P10.1.

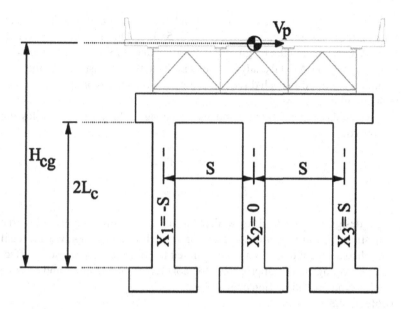

FIGURE P10.2   Problem P10.2.

are given in Figure P10.2 and the ensuing discussion. The clear cover to the transverse bars is 1.5 inches. Specified reinforcing yield strength is $f_y = 60$ ksi (A 706 Grade 60 bars) for both longitudinal and transverse reinforcement. Specified concrete strength is $f'_c = 3$ ksi.

The bridge is a three-span structure consisting of 195-ft end spans and a 230-ft central span, for a total bridge length of 620 feet. The superstructure weight, including live load assumed to be present during strong ground shaking, is 11.8 kips per foot. Seismic design data are as follows:

$$PGA = 0.488 \quad S_S = 0.897 \quad S_1 = 0.227 \quad F_a = 1.050 \quad F_v = 3.32$$

- Column diameter, $D = 54$ inches
- Height to superstructure center of gravity, $H_{CG} = 47.5$ feet
- Clear column height, $2L_C = 33.75$ feet, spacing $S = 16.875$ feet
- Initial column axial loads $P_u = 1,290$ kips (exterior columns), 1,067 kips (center column)
- Displacement demand from response spectrum analysis = 4.64 inches
- Longitudinal bars are 30 #9, #6 hoops spaced at 6 inches

Determine the plastic shear, V_P, and the displacement capacity from a pushover analysis, Δ_{CAP}.

Problem 10.3

For the structure in Problem 10.2, select and scale a set of 14 ground motion records for use in a nonlinear response history analysis of the structure.

The subsurface investigations indicate a Site Class D condition. The modal magnitude, distance pair is M_w, $R=7.75$, 55 km. The fundamental periods are 0.78 seconds in the longitudinal direction and 0.73 seconds in the transverse direction. Modal analysis results indicate that at least 90% of the total mass is captured in each direction at a period of 0.55 seconds.

Problem 10.4

A wide, short-span bridge is modeled as a rigid block with the following properties:

$$W = 2,823 \text{ kips}$$

$$K_i = 178 \text{ kips/inch}$$

The design seismic loading is identical to that for Problem 10.2. Use the substitute structure method to determine a preliminary design yield force if the displacement ductility is to be limited to no more than 6. Use an effective damping model appropriate for concrete substructures, and assume pulse-type ground motion characteristics.

Problem 10.5

A 54-inch diameter column with 20 #10 longitudinal bars (A 706) and #6 hoops at 5 inches on center in the plastic hinging zone has a clear height of 37 feet and behaves as a rigid frame with fixity at both the top and bottom of the column. Specified concrete strength for the column is 4 ksi and the yield strength of the A 706 bars is 60 ksi.

- Determine the overstrength plastic shear of the column. The column axial compression is $P_u=1,221$ kips. Clear cover to the hoops is 2.5 inches.
- Determine the displacement ductility demand on the column if the displacement demand from a seismic analysis of the bridge is 12.6 inches at the pier for which the column is located.
- Determine the displacement ductility capacity of the column in question.
- Determine the adequacy of the column design in terms of seismic shear demand.
- Establish the minimum square footing dimension which will satisfy overturning requirements if the footing depth is 5.5 feet. Check the square footing thus obtained for rigid footing criteria. The nominal bearing resistance is $q_n=20$ ksf.
- Determine whether second-order effects need to be included in the analysis which generate the displacement demand, given as 12.6 inches for this problem. Assume that the dead load, P_{dl}, is 70% of the total load, P_u.

Problem 10.6

A three-span bridge has a total length of 620 feet. One abutment is constructed on multiple rows of piles so as to be extremely stiff in the longitudinal direction. This abutment is considered to be the fixed point on the superstructure. The other abutment incorporates an expansion joint to accommodate seismic and thermal movements. The column height at both piers is 31 feet. The seismic analysis indicates a displacement demand during strong ground shaking of 17.7 inches in the longitudinal direction. Temperature loading (TU) requirements at the expansion abutment necessitate the accommodation of 3.50 inches expansion and 3.50 inches of contraction.

Determine the required seat length at the expansion abutment if the bridge is located in:

- Seismic Design Category A with no skew
- Seismic Design Category B with no skew
- Seismic Design Category C with no skew
- Seismic design Category D with no skew
- Seismic Design Category D with a 45-degree skew of substructures

PROBLEM 10.1	TEH	1/2

CFST24.spf SeismoStruct Project File

- take $f'_{ce} = 1.3 \times 3.0 = 3.9$ ksi, LRFD GS 8.4.4
- take $f_{ye} = 1.5 \times 35 = 52.5$ ksi, expected yield for A53 from LRFD GS 7.3
- take $\lambda_{mo} = 1.2$, overstrength factor from LRFD GS 4.11.2

Apply 117 kip downward loads at each bearing location in the z-direction and a 1-kip incremental load at the center of gravity node in the y-direction and push the pier to failure.

PROBLEM 10.1	TEH	2/2

The plastic shear is found to be 788 kips and the ultimate displacement 8.63 inches. This could be misleading if the strain corresponding to this displacement in the extreme piles is sufficient to cause buckling of the pile wall.

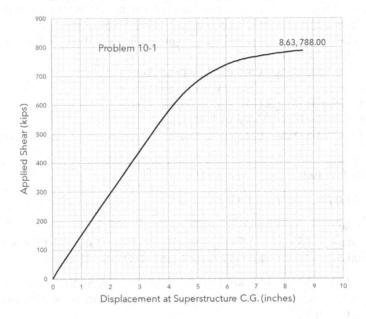

The shear per pile, with *SeismoStruct* overstrength included, is then $1.2 \times (788/9) = 105$ kips. In Chapter 9, Equation 9.94 gives a nominal shear resistance of 386 kips per pile. Hence, shear failure does not preclude flexural failure.

Finally, note that the maximum strain from the model, $\varepsilon = 0.0013$, is significantly less than that corresponding to tube buckling from Chapter 9, Equation 9.96 (0.00977).

The estimated displacement capacity of 8.63 is a realistic estimate since neither shear failure nor tube buckling occurs at this level of loading.

PROBLEM 10.2	TEH	1/4

$S_{DS} = 1.050 \times 0.897 = 0.942$

$S_{D1} = 3.320 \times 0.227 = 0.754$

$T_S = 0.754 / 0.942 = 0.800 \text{ sec}$

$T_0 = 0.20 \times 0.800 = 0.160 \text{ sec}$

Use CONSEC for initial section analysis.

$f'_{ce} = 1.3 \times 3 = 3.90 \text{ ksi}$

$f_{ye} = 68 \text{ ksi} \qquad \varepsilon_{SH} = 0.0125$

$\varepsilon_{ye} = \dfrac{68}{29,000} = 0.002345 \qquad \varepsilon_{su}^{R} = 0.090$

Radius of bar circle $= 27 - 1.5 - .75 - \dfrac{1.128}{2}$

$\qquad = 24.186 \text{ inches}$

$L_c = \frac{1}{2} \times 33.75 = 16.875' = 202.5 \text{ inches}$

$\lambda_{mo} = 1.2$ for A706 reinforcing

Plastic Hinge Length:

$\qquad L_p = 0.08 (202.5) + 0.15 (68)(1.128)$

$\qquad\quad = 16.2'' + 11.5''$

$\qquad\quad = 27.7 \text{ inches} > 2(11.5)$

$\qquad \Rightarrow$ Use $L_p = 27.7$ inches, $L_{sp} = 11.5''$

Section analysis $\Rightarrow$ See p. 4 of 4

PROBLEM 10.2	TEH	2/4

For initial column axial loads:

$P = 1,290 k \Rightarrow \lambda_{mo} M_{pe} = 6,217 ft \cdot k$

$P = 1,067 k \Rightarrow \lambda_{mo} M_{pe} = 5,948 ft \cdot k$

$\Sigma_i d^2 = 2(16.875)^2$
$= 569.53 ft^2$

$H' = H_{cg} - L_c$
$= 30.625'$

$V_1 = \dfrac{6,217}{16.875} \quad V_2 = \dfrac{5,948}{16.875} \quad V_3 = \dfrac{6,217}{16.875}$

$= 368^k \qquad = 352^k \qquad = 368^k$

$\left. \right\} \quad V_{po} = 1,088 \, kips$

$\Delta P_1 = \dfrac{1,088 (30.625)(16.875)}{569.53} = 987 \, kips$

$P_1' = 1,290 - 987 = 303^k \Rightarrow \lambda_{mo} M_{pe} = 5,077$

$P_2' = 1,067 - 0 = 1,067 \Rightarrow \lambda_{mo} M_{pe} = 5,948$

$P_3' = 1,290 + 987 = 2,277 \Rightarrow \lambda_{mo} M_{pe} = 7,172$

$V_1' = 5077 / 16.875 = 301 \, kips$

$V_2' = \qquad\qquad 352 \, kips$

$V_3' = 7,172 / 16.875 = 425 \, kips$

$\underline{V_{po} = 1,078 \, kips}$

$(< 1\% \; change)$

PROBLEM 10.2	TEH	3/4

Take $V_{po} = (1,088 + 1,078)/2$

$V_{po} = 1,083 \text{ KIPS}$

$P_3 = 1,290 + \dfrac{1,083(30.625)(16.875)}{569.53}$

$= 2,273 \text{ KIPS}$

$\Rightarrow \phi_y = 0.0001157 \text{ in}^{-1}$

$\phi_u = 0.0006616 \text{ in}^{-1}$

Δ_y : yield displacement between contraflexure (mid-height of column) and maximum moment (top or bottom of column)

total $\Delta_{yT} = 2\Delta_y$

$\Delta_y = \frac{1}{3} \phi_y (L_c + L_{sp})^2$

$= \frac{1}{3}(0.0001157)(202.5 + 11.5)^2$

$= 1.766 \text{ inches}$

$\Delta_p = (\phi_u - \phi_y) L_p (L_c + L_{sp} - L_p/2)$

$= 0.0005459(27.7)(200.15)$

$= 3.026 \text{ inches}$

$\underline{\Delta_{yT} = 2 \times 1.766 = 3.532 \text{ inches}}$

$\Delta_{pT} = 2 \times 3.026 = 6.052 \text{ inches}$

$\underline{\Delta_{uT} = 3.532 + 6.052 = 9.584 \text{ inches}}$

PROBLEM 10.2	TEH	4/4

Axial Load P, kips	Expected M_{pe}, ft-k	Over-Strength $\lambda_{mo}M_{pe}$, ft-kips	Curvature ϕ_y	Curvature ϕ_u
0	3,874	4,649	0.00009984	0.00113040
250	4,176	5,011	0.00011890	0.00104840
500	4,436	5,323	0.00010891	0.00097376
750	4,682	5,618	0.00010175	0.00090591
1,000	4,957	5,948	0.00012239	0.00085807
1,250	5,181	6,217	0.00011473	0.00081249
1,500	5,391	6,469	0.00010871	0.00076695
1,750	5,585	6,702	0.00010409	0.00072625
2,000	5,793	6,952	0.00012089	0.00069148
2,250	5,960	7,152	0.00011566	0.00066157
2,500	6,112	7,334	0.00011113	0.00063414
2,750	6,246	7,495	0.00010758	0.00060747
3,000	6,391	7,669	0.00012225	0.00058035

PROBLEM 10.3	TEH	1/2

Use SigmaSpectra with target PSA from:

$$S_{DS} = 1.050 \times 0.897 = 0.942$$

$$S_{D1} = 3.320 \times 0.227 = 0.754$$

$$T_s = 0.754 / 0.942 = 0.800 \text{ sec}$$

$$T_0 = .2 \times .800 = 0.160 \text{ sec}$$

Take $T_{upper} = (T_1)_{Max} \times \sqrt{6}$

$$= 0.78 \times \sqrt{6} = 1.91 \text{ sec}$$

$\Rightarrow$ take $T_{upper} = 2$ sec

Take $T_{Lower} = \text{Min} \begin{cases} T_{90} \\ .2(T_1)_{min} \end{cases}$

$T_{90} = 0.55$ sec

$.2(T_1)_{min} = .2 \times 0.73 = .146$

$\Rightarrow$ take $T_{Lower} = 0.14$ sec

$T_{Range} = [0.14, 2.00]$ sec for scaling

Use flat files & spectra data in "GMScaling - GeoMean - 2020.xlsx" to develop a set of 40-50 candidate records for SigmaSpectra. Select no more than 5 candidates from any single event.

Final results → See next page

| PROBLEM 10.3 | | | | | TEH | | | | 2/2 |

EQ	MW	Year	Station	Site Class	SF	MSE	PGA g	PGV cm/s	PGD cm
Maule, Chile	8.80	2010	LACH	C	1.506	0.0109	0.261	32.4	19.0
Landers	7.28	1992	TPPO	D	3.777	0.0126	0.105	15.6	9.2
Cape Mendocino	7.01	1992	FFB	C	3.861	0.0134	0.117	21.4	17.1
Chi-Chi	7.62	1999	TCU067	C	1.015	0.0134	0.425	73.7	78.7
Taiwan SMART1(45)	7.30	1986	M12	D	2.781	0.0180	0.143	24.5	9.7
Taiwan SMART1(45)	7.30	1986	M11	D	2.773	0.0187	0.125	24.9	9.9
Landers	7.28	1992	AMBOY	C	3.682	0.0192	0.131	18.0	8.8
Kocaeli	7.51	1999	YARIMCA	D	1.594	0.0194	0.286	70.9	63.1
Chi-Chi	7.62	1999	CHY015	D	2.456	0.0211	0.160	25.2	11.1
Chi-Chi	7.62	1999	TCU122	C	2.162	0.0223	0.228	42.6	40.2
Darfield	7.00	2010	CCBG	D	2.697	0.0238	0.159	44.9	40.3
Taiwan SMART1(45)	7.30	1986	M01	D	2.688	0.0244	0.139	24.0	10.8
Darfield	7.00	2010	GDLC	D	0.727	0.0246	0.731	117.1	77.6
Taiwan SMART1(45)	7.30	1986	O09	D	3.544	0.0250	0.101	14.5	8.3

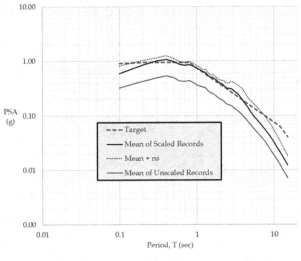

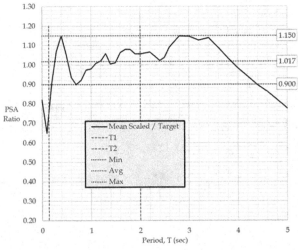

PROBLEM 10.4	TEH	1/1

With $\mu_D = 6$:

$$\xi_{EFF} = 0.05 + 0.444 \left(\frac{5}{6\pi} \right)$$

$$\Rightarrow \xi_{EFF} = 0.1678 \; (16.8\%)$$

Pulse-Type $R_\xi = \left(\frac{.07}{.02 + .1678} \right)^{0.25}$

$$\Rightarrow R_\xi = 0.7814$$

$$T_i = 2\pi \sqrt{\frac{2,823}{386 \times 178}} = 1.274 \, sec$$

$$T_{EFF} = 1.274 \sqrt{6} = 3.12 \, sec$$

$$PSA_{TEFF} = \frac{S_{D1}}{T_{EFF}} = \frac{3.32 \times 0.227}{3.12}$$

$$\Rightarrow PSA_{TEFF} = 0.241$$

$$D_{Max} = 0.241 \times 386 \times 0.7814 \left(\frac{3.12}{2\pi} \right)^2$$

$$\Rightarrow D_{Max} = 17.96 \, inches$$

$$D_y = 17.96 / 6 = 2.994 \, inches$$

$$\left(F_y \right)_{Reqd} = 178 \frac{K}{in} \times 2.994 \, in.$$

$$\Rightarrow \left(F_y \right)_{Reqd} = 533 \, kips$$

$$(= 0.189 \, W)$$

PROBLEM 10.5	TEH	1/5

Expected Strengths:

$$f'_{ce} = 1.3 \times 4 = 5.2 \text{ ksi}$$

$$f_{ye} = 68 \text{ ksi} \qquad f_{ue} = 95 \text{ ksi}$$

A706 #10 Bar Stress-Strain:

$$\varepsilon_{sh} = 0.0115$$

$$\varepsilon_{ye} = 0.0023$$

$$\varepsilon_{su}^R = 0.090$$

Equations From Section 9.7:

$$\phi_y \cong 2.25 (0.0023)/54$$

$$= 0.0000958$$

$$d_s = 54'' - 2 \times 2.5'' - 6/8$$

$$= 48.25''$$

$$P_r = \frac{4 \times 0.44}{48.25 \times 5} = 0.0073$$

$$A_c = \pi (48.25)^2/4 = 1,828 \text{ in}^2$$

$$A_s = 20 \times 1.27 = 25.4 \text{ in}^2$$

$$s' = s - 0.75 = 4.25''$$

$$P_{cc} = 25.4/1,828 = 0.0139$$

$$k_e = \frac{\left(1 - \frac{4.25}{2 \times 48.25}\right)^2}{1 - 0.0139} = 0.927$$

Use $f_{yh} = 60$ ksi for the hoops.

$$f'_x = \frac{1}{2}(0.927)(0.0073)(60)$$

$$= 0.203 \text{ ksi}$$

$$f'_{cc} = 5.2(-1.254 + 2.580 - 0.078)$$

$$= 1.248(5.2)$$

$$f'_{cc} = 6.49 \text{ ksi}$$

$$\varepsilon_{cu} = 0.004 + \frac{1.4(.0073)(60)(.09)}{6.49}$$

$$\varepsilon_{cu} = 0.0125$$

$$\frac{c}{D} \cong 0.2 + 0.65\left(\frac{1221}{5.2 \times 2290}\right)$$

$$= 0.267$$

$$c \cong 0.267 \times 54 = 14.4''$$

$$d = 54 - 2\frac{1}{2} - .75 - 1.27/2 = 50.1''$$

$$\phi_u = \text{Min} \begin{cases} 0.0125/14.4 \\ 0.09/(50.1-14.4) \end{cases}$$

$$= \text{Min} \begin{cases} 0.00087 \\ 0.00252 \end{cases}$$

$$\Rightarrow \phi_u = 0.00087$$

$$L_{sp} = 0.15(68)(1.27) = 12.95''$$

$$L_c = \frac{1}{2} \times 37' = 18.5' = 222''$$

(contraflexure at mid-height
      for rigid frame behavior)

PROBLEM 10.5	TEH	3/5

$$0.08 L_c = 0.08 \times 222 = 17.76''$$

$$L_p = 17.76'' + 12.95'' = 30.71''$$

$$L_c + L_{sp} - L_p/2 = 220''$$

$$L_c + L_{sp} = 235''$$

$$\Delta_y = \tfrac{1}{3}(0.0000958)(235)^2$$

$$= 1.76'' \quad \text{from mid-height} \\ \text{to top \& from} \\ \text{mid-height to base}$$

$$(\Delta_y)_{TOT} = 2 \times 1.76 = 3.52''$$

$$\Delta_p = (.00087 - .0000958)(30.71)(220)$$

$$= 5.23''$$

$$(\Delta_p)_{TOT} = 2 \times 5.23 = 10.46''$$

$$\Delta_u = 3.52 + 10.46 = 13.99''$$

CONSEC gives $\Delta_u = 13.00''$ but does not account for strain penetration. Also from CONSEC:

$$M_{pe} = 4,773 \, ft \cdot k$$

$$M_{po} = 1.2 \times 4,773 = 5,728 \, ft \cdot k$$

$$\underline{V_{po} = \frac{5,728}{37 \times \frac{1}{2}} = \underline{310 \, KIPS}}$$

$$\mu_{DEM} = 12.6''/3.52'' = 3.58$$

$$\mu_{CAP} = 13.99''/3.52'' = 3.97$$

PROBLEM 10.5	TEH	4/5

Shear Resistance in the Hinging Zone of the Column.

$f_s = 0.0073 \times 60 = 0.438 > 0.35$

$\Rightarrow$ use $f_s: 0.35$

$\alpha' = \dfrac{0.35}{0.15} + 3.67 - 3.58 = 2.42$

$1 + \dfrac{P_u}{2A_g} = 1 + \dfrac{1,221}{2 \times 2,290} = 1.27$

$\dfrac{\upsilon_c}{\sqrt{f'_c}} = Min \begin{cases} 0.032(2.42)(1.27) \\ 0.110 \\ 0.047(2.42) \end{cases}$

$\dfrac{\upsilon_c}{\sqrt{f'_c}} = Min \begin{cases} 0.0983 \\ 0.1100 \\ 0.1137 \end{cases} = 0.0983$

$\upsilon_c = 0.0983 \sqrt{4} = 0.197 \, ksi$

$V_c = 0.197 (0.80 \times 2,290) = 361 \, kips$

(some would use $f'_{ce} = 5.2 \, ksi$ since this was used for section analysis; using $f'_c = 4$ is conservative)

$V_s = \dfrac{\pi}{2} \left(\dfrac{0.44 \times 60 \times 48.25}{5} \right) = 400 \, kips$

$(V_s)_{max} = 0.25 \sqrt{4} \, (.8 \times 2,290) = 916 \, kips$

$\Rightarrow V_s = 400 \, kips$

$\phi(V_c + V_s) = 0.90(361 + 400)$
$\qquad = 685 \, kips > V_{po} = 310^k, \; Shear \; ok$

Footing Analysis

$$M_{po} + V_{po} H_f = 5,728 + 310(5.5)$$
$$= 7,433 \text{ FT·KIPS}$$

Square Footing $\Rightarrow$ $L = B$

$$P_u \left(\frac{L-a}{2} \right) = P_u \left(\frac{B}{2} - \frac{P_u}{2 q_n B} \right)$$

$$1,221 \left(\frac{B}{2} - \frac{1,221}{2(20)B} \right) \geq 7,433$$

$$\Rightarrow B \geq 15.99 \text{ FT}$$

$\Rightarrow$ $\underline{\text{USE } 16' \times 16' \text{ Footing}}$

$$\frac{L - D_c}{2 H_f} = \frac{16 - 4.5}{2 \times 5.5} = 1.045 < 2.5$$
$$\Rightarrow \text{OK (Rigid)}$$

P-Δ Assessment

$$P_{dL} = 0.7 \times 1,221 = 855 \text{ KIPS}$$
$$\Delta_r = \frac{1}{2} \Delta_{DEM} = \frac{1}{2} \times 12.6 = 6.3''$$
$$M_p = M_{pe} = 4,773 \text{ FT·K}$$
$$P_{dL} \Delta_r = 855 \times 6.3 / 12 = 449 \text{ FT·K}$$
$$0.25 M_p = 0.25 \times 4,773 = 1,193 \text{ FT·K}$$

$449 < 1,193$ $\Rightarrow$ No need to have
included 2nd-order
effects in the
analysis for
displacement demand.

PROBLEM 10.6	TEH	1/1

$L = 620 \text{ ft} \qquad H = 31'$

<u>Seismic Design Category A</u>

$N \geqslant 8 + .02 \times 620 + .08 \times 31$

$\qquad = 22.9 \text{ inches}$

<u>Seismic Design Category B (or c)</u>

$N \geqslant 1.5 \times 22.9 = 34.3''$

<u>Seismic Design Category D</u>

$N \geqslant 4 + 1.65 \times 17.7 = 33.2''$

<u>Seismic Design Category D, 45° skew</u>

$N \geqslant 33.2 \left(1 + 0.00025 \times 45^2\right)$

$\qquad = 50.0''$

Note: SDC A: $N \geqslant 17.7 + 3.5 = 21.2''$
$\qquad\qquad\qquad$ from analysis
$\qquad\qquad \Rightarrow \text{use } N \geqslant 22.9''$

$\qquad$ SDC B,C: $N \geqslant 17.7 + 3.5 = 21.2''$
$\qquad\qquad\qquad$ from analysis
$\qquad\qquad \Rightarrow \text{use } N \geqslant 34.3''$

$\qquad$ SDC D: $N \geqslant 33.2 + 3.5$
$\qquad$ (No skew)
$\qquad\qquad \Rightarrow N \geqslant 36.7''$

$\qquad$ SDC D: $N \geqslant \qquad + 3.5$
$\qquad$ (skew)
$\qquad\qquad \Rightarrow N \geqslant 53.5''$

10.8   EXERCISES

E10.1.

For the prestressed girder alternative of the Project Bridge, estimate the displacement capacity of the pier in the transverse direction (rigid frame behavior) given the following data. Use hand calculations and base the results on the initial column axial loads.

- column diameter = 42 inches
- clear column height = 29 feet
- distance from column base to superstructure center of gravity = 39 feet
- two-post pier, column spacing $S = 18.5$ feet
- $f'_c = 3$ ksi
- $f_y = f_{yh} = 60$ ksi, A 706 bars
- longitudinal bars (per column), 14 #9
- transverse hoops (per column), #5 at 4 inches on center
- clear cover to hoops = 2 inches
- initial column loads (per column), $P_u = 450$ kips

E10.2.

For the prestressed girder alternate of the Project Bridge, estimate the expected, overstrength plastic shear of the pier. See Exercise E10.1 for details of the pier. Revise the displacement capacity estimated in E10.1 to account for the increased column axial loads during strong ground shaking.

E10.3.

For the prestressed, concrete girder alternative of the Project Bridge, estimate the shear resistance of the columns for the expected, overstrength plastic shear seismic loading calculated in E10.2. See Exercises E10.1 and E10.2 for additional details.

E10.4.

Suppose the project bridge is located at a site with the seismic data below. The superstructure weight is estimated to be 9.4 kips per linear foot for the concrete girder option, including two lanes of uniform live load. The column weights are 1.44 kips per linear foot. The pier cap and pier diaphragm together weigh 145 kips. Abutment weights are 94 kips each, including the abutment beam and backwall. Each integral, stub abutment is supported by six piles with an estimated lateral stiffness of 20 kip per inch per pile. The two 42-inch diameter columns at the pier are assumed to behave as cantilevers in the longitudinal direction with clear height equal to 29 feet. No passive resistance at the abutments is to be relied upon for seismic resistance.

a) Determine the seismic displacement in the longitudinal direction using a simplified linear response spectrum analysis. Apply the short-period displacement amplification, R_d, if applicable (see Equation 10.5). Cracked properties of the concrete columns may be taken to be equal to 35% of gross properties.

b) Determine the longitudinal seismic shear demand on the pier columns from the simplified response spectrum analysis using a response modification factor $R = 3.5$.

c) Using an estimated yield curvature, $\phi_y = 0.000111$ in^{-1}, and ultimate curvature, $\phi_u = 0.001390$ in^{-1}, estimate yield and ultimate displacement for the pier in the longitudinal direction. Other pertinent data may be taken from Exercise 10.1.

d) Determine the ductility demand on the pier in the longitudinal direction.

e) Using an appropriate bi-linear load-deflection relationship, apply a substitute-structure method analysis to obtain a second estimate of the longitudinal displacement. Assume that the abutments remain linear and experience no damage.

f) Given that the abutments have been assumed to remain in the linear elastic range of behavior, determine the required shear resistance for abutment piles.

g) Select and scale a 14-record ground motion suite appropriate for use in a non-linear response history analysis of the Project Bridge using the design response spectrum as the target response spectrum with a target standard deviation (natural logarithm-based) of 0.6 across all periods. The modal magnitude for the project site is $M_w = 7.6$. Near-field records are not considered appropriate for the Site Class D project location. Use the natural period computed in the previous step to establish a rational period range of interest for ground motion scaling.

$$A_S = 0.644$$

$$S_{DS} = 0.927$$

$$S_{D1} = 0.863$$

$$T_L = 8 \text{ seconds}$$

11 Seismic Isolation of Bridges

Properties of isolation bearings, including both lead-rubber bearings (LRB) and friction-pendulum systems (FPS) were discussed and summarized in Section 8.3.

In the current chapter, preliminary design considerations, based on partial isolation, are presented for a constructed bridge on Interstate 40 over State Route 5 in Madison County, Tennessee. In partial isolation for this structure, isolation devices were incorporated into the pier, with abutments remaining integral. Also included is the ground motion selection and modification which was adopted for the final design of the bearings. The preliminary design using the substitute-structure method (SSM) presented here (see Section 10.5) resulted in bearing specifications which required no modification when the inelastic response history analysis was performed. That is to say, the simplified, hand-calculated solution using a modified substitute-structure type of analysis, accurately predicted inelastic response history results.

Also presented are simple, preliminary calculations for the Hernando De Soto Bridge carrying Interstate 40 over the Mississippi River in Memphis, Tennessee. The arch bridge has been retrofitted with FPS devices.

11.1 PARTIAL ISOLATION OF INTERSTATE 40 OVER STATE ROUTE 5

Modern seismic design of bridges typically requires that substructures be designed for the over-strength plastic shear of the columns. For the subject bridge, over-sized piers were used for aesthetic reasons. Design of foundations and pier caps for the plastic shear of over-sized columns can require dramatic cost increases for the substructures – piling, pile caps, shafts, etc. With continued emphasis upon the seismic hazard in the New Madrid Seismic Zone (NMSZ), non-traditional techniques, like isolation, may need to be considered in an increasing number of cases. For this design, a three-span bridge in Madison County, Tennessee was analyzed for both non-isolated and partially isolated conditions with the goal of reducing the demand on piling and pile caps to at least partially offset the cost of the isolators. Partial isolation offers the benefit of elastic substructures at the piers without costly expansion joints at the abutments. The final design for this structure included the partial isolation alternative.

The subject bridge consists of three continuous, steel I-girder spans (95 ft, 156 ft, 95 ft long) and is 129.25 feet wide. The superstructure consists of fifteen welded steel plate girders and a composite cast-in-place concrete deck with 3 ft high traffic parapets and a design allowance for a 3.5 inch asphalt overlay. The total superstructure

DOI: 10.1201/9781003265467-11

weight is 28.6 kips per foot with a center of gravity 4.50 feet above the top of the pier cap. Integral abutments on friction pipe piles are located at each end of the bridge. All substructures are skewed 4.9 degrees from normal. An isometric model of the structure is given in Figure 11.1. The cross section of the bridge is shown in Figure 11.2.

Table 11.1 summarizes seismic design data for the design basis event (DBE) and for the maximum considered event (MCE). The hazard levels are taken to correspond to the 7% probability of exceedance (PE) in 75 years and the 3% probability of exceedance (PE) in 75 years, respectively. Whereas the 7% PE in 75-years ground shaking is specified for design in the LRFD BDS, the 3% PE in 75-years ground shaking was used for design of the isolation system on this project.

Passive resistance from backfill at the abutments was conservatively neglected for both non-isolated and isolated conditions. Abutment piles were specified as 20 inch diameter × 5/8 inches thick since anticipated displacement demands were greater than 4 inches. The LRFD GS, in Section 5.2.4.2, requires that smaller diameter piles be ignored for lateral stiffness in such cases:

> "For pile-supported abutment foundations, the stiffness contribution of piles less than or equal to 18 in. in diameter or width shall be ignored if the abutment displacement is greater than 4 in., unless a displacement capacity analysis of the piles is performed, and the piles are shown to be capable of accommodating the demands."

The abutment pile cross-section D/t-ratio qualifies as ductile. This was judged necessary to prevent local buckling of the abutment piles during strong ground shaking.

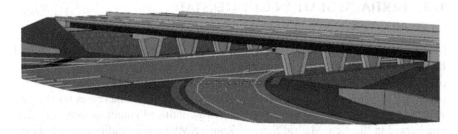

FIGURE 11.1   Interstate 40 over State Route 5 – Madison County, Tennessee.

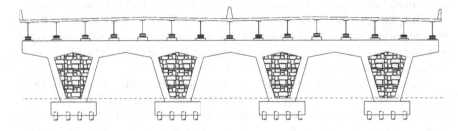

FIGURE 11.2   Interstate 40 cross section.

TABLE 11.1

Seismic Design Parameters for Interstate 40 over State Route 5

Parameter	DBE Hazard-level	MCE Hazard-level
PGA	0.279	0.482
S_S	0.554	0.952
S_I	0.148	0.269
A_S	0.347	0.491
S_{DS}	0.626	0.917
S_{DI}	0.411	0.655
T_S	0.656	0.714
T_O	0.131	0.143
$T*$	0.820	0.893
Latitude	35°40'14" N	35°40'14" N
Longitude	88°49'47" W	88°49'47" W
Site Class	D	D

The natural periods of the non-isolated bridge are 0.21 seconds (transverse direction) and 0.76 seconds (longitudinal direction). The pier design basis, for either the 1,000-yr or the 2,500-yr ground motion, is the over-strength plastic shear, V_{PO}. For even the minimum amount of reinforcement in the columns, V_{PO} was estimated to be 8,044 kips per pier. This is the problematic feature of the subject bridge – due to the width of the structure combined with the over-sized columns, even the minimum reinforcement permitted by the LRFD BDS results in extremely large seismic shears. The pile caps and piling must be designed to remain elastic when subjected to the large over-strength shears and moments at the base of the columns, for a design without the isolation bearings.

Ground motion record pairs were selected from appropriate magnitude events and Site Class D or E conditions at the recording station. Each record pair was scaled to match the target spectrum in the period range of 0.20–4.00 seconds by minimizing the mean square error between each individual record pair and the MCE target spectrum. These record pairs were then used for the nonlinear response history analysis (NLRHA) of the isolated bridge. Figure 11.3 includes the plot of the ground motion average spectra for comparison to the target spectrum at the MCE-level of ground shaking. Table 11.2 summarizes the 14-record suite. A 14-record suite is twice the minimum of seven required in AASHTO. Four of the records (nos. 11–14) are synthetic records developed in published research specifically for the NMSZ (Atkinson and Beresnev, 2002, Fernández, 2007). Since only the mean response was to be determined, no target, log-based standard deviation was specified. This would likely be a needed change for future performance-based design procedures.

For the selected record suite, the minimum suite mean-to-target PSA ratio in the period range of interest is 0.913 (greater than 0.900), the average ratio is 1.047 (greater than 1.000), and the maximum ratio is 1.177 (less than 1.300). The ground

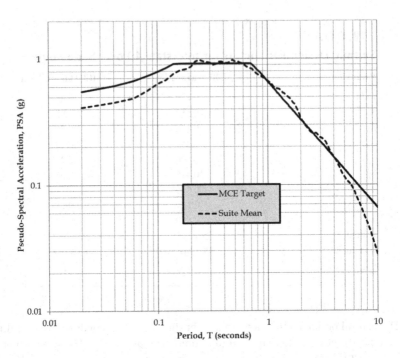

FIGURE 11.3 Target and suite mean PSA.

TABLE 11.2
Ground Motion Record Metadata

No.	Event	Year	Station	M_w	R km	V_{S30} m/s	Site Class	f
1	Tabas, Iran	1978	Boshrooyeh	7.35	24.1	338	D	3.84
2	Taiwan SMART1(45)	1986	I01	7.30	56.0	274	D	2.37
3	Landers	1992	Amboy	7.28	69.0	271	D	3.01
4	Landers	1992	Yermo FS	7.28	23.6	354	D	2.18
5	Kocaeli, Turkey	1999	Bursa Tofas	7.51	60.0	275	D	3.55
6	Chi-Chi, Taiwan	1999	CHY036	7.62	16.1	233	D	1.49
7	Chi-Chi, Taiwan	1999	HWA048	7.62	47.4	278	D	2.56
8	Sierra el Mayor	2010	M5057	7.20	41.0	163	E	2.32
9	Darfield, NZ	2010	HORC	7.10	7.0	326	D	0.96
10	Darfield, NZ	2010	REHS	7.10	19.0	141	E	1.67
11	Atkinson-Beresnev	2001	Memphis	8.00	53.8	205	D	1.44
12	GaTech-Fernández	2006	Paducah	7.70	25.9	295	D	1.35
13	GaTech-Fernández	2006	Paducah	7.70	25.9	295	D	1.18
14	GaTech-Fernández	2006	Jonesboro	7.70	30.4	205	D	1.55

Note: Records 11–14 are synthetic records from the literature.

motion suite was thus deemed appropriate for design of the isolators based on mean response for the inelastic response history analyses (14 record pairs; 14 demand values).

Table 11.3 is a summary of the ground motion parameters corresponding to the 14-record-pair suite.

Effective damping and stiffness properties were adopted for a substitute-structure-method (SSM), simplified response spectrum analysis. This is the method first proposed by Gulkan and Sozen (Gulkan and Sozen, 1974), further developed by Priestley and others (Priestley et al., 2007), and the basis for the simplified analysis procedures in the AASHTO GS ISO (AASHTO, 2014).

Effective stiffness was taken to be equal to the secant stiffness. Effective damping is imparted to the system through hysteretic behavior of yielding elements. An isolation device may be completely defined by the three parameters: (a) characteristic strength, Q_d, (b) post-yield stiffness, K_d, and (c) post-yield stiffness ratio, $\alpha = K_d/K_i$, where K_i is the initial stiffness of the isolator. The isolator yield displacement, D_y, is related to α, Q_d, and K_d as given by Equation 11.1.

For the analysis of the LRB isolators on this project, α was taken to be equal to 0.10. The initial stiffness is due to the lead plug and elastomer stiffness values in parallel with one another. Once the lead plug has yielded in shear, only the stiffness contribution from the elastomer remains. The true yield strength is related to the characteristic strength by Equation 11.2.

Rather than computing the effective stiffness of the isolators at a given pier, it is convenient to recognize that the isolator-pier assembly in series results in a bilinear behavior as depicted in Figure 11.4. The composite stiffness values, K_{SUB1} and K_{SUB2}, depend on the isolator parameters – K_d and α – and on the pier stiffness, K_{PIER}, and can be shown to be given by Equations 11.3 and 11.4.

TABLE 11.3
Suite Ground Motion Parameters

Parameter	MCE Record Set
Max acceleration (g)	0.397
Max velocity (cm/sec)	62.9
Max displacement (cm)	36.1
$5 \times V_{max}/A_{max}$ (sec)	0.81
$8 \times D_{max}/V_{max}$ (sec)	6.89
Acceleration RMS (g)	0.063
Velocity RMS (cm/sec)	13.5
Displacement RMS (cm)	10.0
Arias intensity (m/sec)	4.03
Specific energy density (cm²/sec)	12,035
Cum. abs. velocity (cm/sec)	2,412
Acc spectrum intensity (g-sec)	0.364

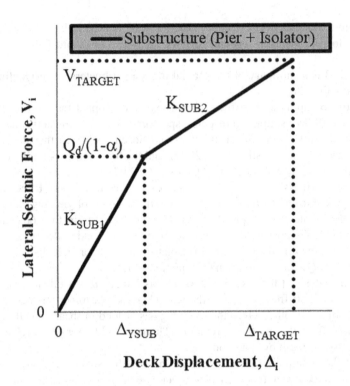

FIGURE 11.4 Bi-linear load displacement.

The real power of this direct displacement-based design procedure is apparent in that the engineer decides upfront the desired values for V_{TARGET} and Δ_{TARGET}, the shear and displacement, respectively, at each pier under seismic loading.

It is suggested here that isolators at each pier be designed such that V_{TARGET} and Δ_{TARGET} are identical for each pier, resulting in a rigid body translation of the deck with minimal torsional effects from eccentricity of mass with respect to the center of strength or stiffness. One-half of the smallest yield shear for any of the piers is suggested as an approximate value for V_{TARGET}. Higher mode effects from pier local vibration modes will increase the total shear on the pier and are, in many cases, of the same order as the isolation mode shear, V_{TARGET}; hence, the reason for starting with such a small value for V_{TARGET}. It will be convenient to express Q_d in terms of the other variables. Given target shears and displacements and preliminary values for isolator post-yield parameters, the resulting expression is given by Equation 11.5. For cases in which it is more desirable to assign a preliminary value for Q_d and calculate the necessary K_d, Equation 11.6 is provided.

Composite values for post-yield stiffness ratio, α_{SUB}, and displacement ductility, μ_{SUB}, are given in Equations 11.7 and 11.8, respectively, and are used to compute the effective damping at each substructure. This is the slight deviation from the procedure described in AASHTO (AASHTO, 2014). The effective damping (ξ_{EFF}) for each pier is determined from Equation 11.9. Since a uniform displacement profile is targeted, the total system damping is computed from Equation 11.10. The damping response modifier ($B_L = 1/R_\xi$) is given by Equation 11.11. In Equation 11.10, K_{ABUT} is the elastic stiffness of the abutment, taken to be equal to the abutment pile stiffness only for this project, with no contribution from backfill. The suggested value for ξ_{ABUT}, the elastic damping of the abutments, is 0.05.

It is further assumed that the design response spectrum is inversely proportional to the effective period so that Equations 11.12 and 11.13 are valid. In other words, the assumption is made that the effective period is between T_S and T_L, with verification after the fact. This will typically be the case.

Here, SD is the spectral displacement (inches) at T_{EFF}, SA is the spectral acceleration (g) at T_{EFF}, and S_{D1} is the spectral acceleration (g) at a period of 1 second. The spectral displacement thus computed is the total displacement of the mass. This is compared to an initially assumed displacement. Once the desired accuracy has been attained, the iterations are stopped and the design displacement has been determined.

A target, unidirectional deck displacement of 6 inches was selected to derive trial isolator properties using the simplified approach. Upon iteration for the MCE-level response spectrum, the converged deck displacement was found to be 5.60 inches. The characteristic strength was selected to ensure that the isolator lead plugs do not yield under Strength limit state loads (wind, braking, live, thermal, etc.). The simplified analysis is a unidirectional analysis. Bidirectional effects were approximated in the simplified analysis by assuming:

1. The maximum component is equal to 1.3 times the geometric mean of two components and
2. When one component (transverse or longitudinal) is at its maximum, the perpendicular component is at 60% of its maximum.

This is a departure from the practice typically taken in design offices today. Justification for the method may be found in the literature (Huff, 2016b). The bidirectional displacement demand from the simplified analysis was determined to be 8.26 inches, including thermal offset effects. These required properties of the isolators from preliminary design are listed in Table 11.4. Multiple LRB manufacturers'

TABLE 11.4

Preliminary LRB Isolator Design from Simplified Analysis

Substructure	Q_d k/beam	k_d k/in/beam
Abutment 1	Integral	Integral
Pier 1	15.8	5.7
Pier 2	15.8	5.7
Abutment 2	Integral	Integral

catalogs (FIP Industriale, 2011, Maurer, 2011, Robinson Seismic Ltd., 2011) provide isolators capable of displacing more than 8.26 inches while carrying vertical loads comparable to those expected at the piers of the subject structure.

Preliminary design with extremely stiff substructures may be carried out following procedures for the SSM outlined in Chapter 10, Section 10.4.

$$D_y = \frac{Q_d}{K_d} \cdot \frac{\alpha}{1-\alpha} \qquad (11.1)$$

$$F_y = \frac{Q_d}{1-\alpha} \qquad (11.2)$$

$$K_{SUB1} = \frac{K_{PIER}K_d}{\alpha K_{PIER} + K_d} \qquad (11.3)$$

$$K_{SUB2} = \frac{K_{PIER}K_d}{K_{PIER} + K_d} \qquad (11.4)$$

$$Q_d = V_{TARGET}\left(1 + \frac{K_d}{K_{PIER}}\right) - \Delta_{TARGET}K_d \qquad (11.5)$$

$$K_d = \frac{V_{TARGET} - Q_d}{\Delta_{TARGET} - \dfrac{V_{TARGET}}{K_{PIER}}} \qquad (11.6)$$

$$\alpha_{SUB} = \frac{\alpha K_{PIER} + K_d}{K_{PIER} + K_d} \qquad (11.7)$$

$$\mu_{SUB} = \frac{\Delta_{TARGET}}{\Delta_{YSUB}} = \frac{\Delta_{TARGET}\left(1-\alpha\right)K_{PIER}K_d}{Q_d\left(\alpha K_{PIER} + K_d\right)} \qquad (11.8)$$

$$\xi_{SUB} = \frac{2\left(\mu_{SUB} - 1\right)\left(1 - \alpha_{SUB}\right)}{\pi\mu_{SUB}\left(1 + \alpha_{SUB}\mu_{SUB} - \alpha_{SUB}\right)} \qquad (11.9)$$

$$\xi_{EFF} = \frac{2K_{ABUT}\xi_{ABUT} + \displaystyle\sum_{PIERS}\left(\xi_{SUB}\dfrac{V_{TARGET}}{\Delta_{TARGET}}\right)}{2K_{ABUT} + \displaystyle\sum_{PIERS}\left(\dfrac{V_{TARGET}}{\Delta_{TARGET}}\right)} \qquad (11.10)$$

$$B_L = \frac{1}{R_\xi} = \left(\frac{\xi_{EFF}}{0.05}\right)^{0.30} \leq 1.7 \tag{11.11}$$

$$T_{EFF} = 2\pi \sqrt{\frac{W}{g\left[2K_{ABUT} + \displaystyle\sum_{PIERS}\left(\frac{V_{TARGET}}{\Delta_{TARGET}}\right)\right]}} \tag{11.12}$$

$$SD = \frac{S_{D1}}{T_{EFF}} \cdot \frac{g}{B_L} \cdot \left(\frac{T_{EFF}}{2\pi}\right)^2 = \frac{g}{4\pi^2} \cdot \frac{S_{D1}}{B_L} \cdot T_{EFF} \tag{11.13}$$

$$\xi^* = \xi \frac{1 - 0.1(\mu-1)(1-\alpha)}{\sqrt{\dfrac{\mu}{1+\alpha\mu-\alpha}}} \tag{11.14}$$

Nonlinear modal response history analysis (NLRHA) – Fast Nonlinear Analysis (FNA) –methods were used to verify the simplified SSM design. These procedures are extremely efficient and accurate for problems in which nonlinear behavior is confined to link-type elements. CSiBridge (Computers and Structures, Inc., 2016) was used for the analysis of the bridge. Nonlinear link elements in the CSiBridge library include biaxial hysteretic elements with coupled plasticity for the two shear deformations. The model incorporated into CSiBridge is that proposed by Wen (Wen, 1976) and recommended for isolators by Tsopelas et al. (1991). The isolators were modeled with non-linear properties in both shear directions.

To capture the most accurate load transfer into the non-linear link elements, a finite element model of the bridge was created. The bridge superstructure was modeled using a mixed use of frame and shell elements to best represent the components of the welded plate girder and composite slab. Pipe pile supports at the abutments were modeled as elastic springs at each girder having full vertical restraint. The piers were modeled as non-prismatic frame elements with all column bases fixed for translation and rotation. The girder bases were connected to the pier cap using the non-linear links defined for the isolators.

Ritz vectors were used instead of the usual eigenvector modal analysis. Ritz vectors are generally preferred for this type of analysis (FNA) unless every possible mode is included in the eigenvector analysis.

Elastic damping was taken as 2.5% of critical for the response history analyses. This small damping value was adopted due to problems which can frequently occur with viscous damping models in nonlinear analyses. It has been shown (Priestley and Grant, 2005) that the appropriate elastic damping for nonlinear analyses is significantly less than the typical assumed value of 5%. Suggested values (Priestley et al., 2007) developed in the referenced works are repeated here in Equation 11.14.

For LRB systems, $\alpha = 0.10$ and the above becomes zero when $\mu = 12.111$. So, at least for the case of Rayleigh damping, it would appear that the choice of damping less than 5% of critical is appropriate for the nonlinear analysis of isolated structures given that it is not at all uncommon for isolated systems to have μ values well above 12.

Final design isolator demands – taken as the average from the 14 NLRHA results at each hazard level – are summarized in Table 11.5.

The correct means of calculating the isolator demand for a load case is to compute the resultant displacement at each time step and find the maximum for that case. Undue conservatism would result if the maximum longitudinal demand were combined with the maximum transverse demand since the two do not generally occur simultaneously. Clearly, the simplified preliminary design procedure proved reliable in this case. This can be expected when the deck aspect ratio is such that the superstructure behaves very nearly as a rigid block.

A key parameter in determining the economy of an isolation system for bridges is the shear transmitted to the substructures. Table 11.6 summarizes these values for both simplified analysis and the NLRHA.

Partial isolation resulted in a significant decrease (89% decrease, relative to overstrength plastic shear) in the load transmitted to the substructures at the intermediate piers, compared to the non-isolated conventional design strategy. This does carry along an increase in the load transferred to the abutment piling. However, the nominal shear resistance of the 20 inch diameter × 5/8 inch thick pipe piles with a minimum yield strength of 35 ksi is 356 kips per pile. The load transmitted to the abutments during the 2,500-yr ground shaking was estimated to be 155 kips per pile, about 44% of the nominal capacity.

TABLE 11.5
Isolator Displacement Demands

Substructure	SSM MCE TDD inches	NLRHA MCE TDD inches
Pier 1	8.02	7.85
Pier 2	8.02	7.85

TABLE 11.6
Shear Transmitted to Piers

Condition	SSM kips/pier	NLRHA kips/pier
Non-isolated	8,044	8,044
Isolated	1,185	857

11.2   SEISMIC RETROFIT OF INTERSTATE 40
OVER THE MISSISSIPPI RIVER

The Hernando de Soto Bridge carries Interstate 40 across the Mississippi River near Memphis, Tennessee. The bridge consists of two 900-ft tied arches across the river with multiple approach spans on both sides of the main spans.

The bridge has been retrofitted with friction-pendulum system devices at the three arch piers and LRB isolation bearings at approach spans.

Problem 11.1 describes some of the bearing details and the seismic analysis results for the FPS devices.

11.3   SOLVED EXAMPLES

Problem 11.1

Properties for the friction-pendulum system bearings on Interstate 40 over the Mississippi River are summarized below. Figures P11.1a and P11.1b depict the design response spectra and the arch spans elevation, respectively. Perform a simple, substitute-structure analysis of the isolated bridge as a sanity check on the reported isolator demands. Refer to Chapter 8, Section 8.3, for equations defining the FPS properties. The SSM analysis is discussed in Section 10.4 of Chapter 10. The deck width is 88 feet.

End Piers A and C:

- $R = 244$ inches, radius of concave surface
- $\mu = 0.06$, dynamic friction coefficient
- $D_{Max} = 27.25$ inches, from inelastic response history analysis
- $W = 5,405$ kips per bearing (2 bearings per Pier)
- $D = 8$ ft 10 inches, diameter of bearing

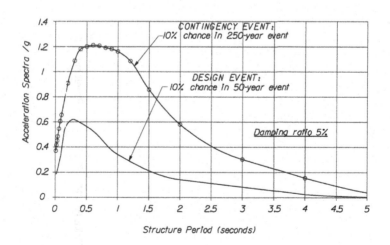

FIGURE P11.1A   DRS for Problem P11.1.

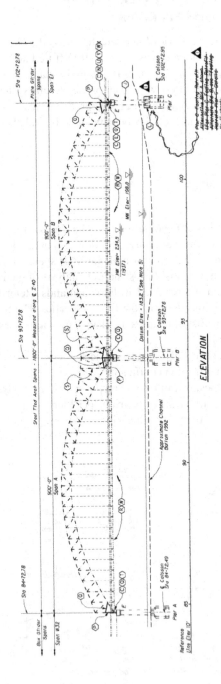

FIGURE P11.1B   Elevation for Problem P11.1.

Center Pier B:
- $R = 244$ inches, radius of concave surface
- $\mu = 0.06$, dynamic friction coefficient
- $D_{Max} = 18.75$ inches, from inelastic response history analysis
- $W = 12{,}611$ kips per bearing (two bearings per pier)
- $D = 8$ ft 10 inches, diameter of bearing

Estimate the maximum force exerted on the piers and the vertical displacements as well.

Assume that the given design response spectrum is a geometric mean-based design response spectrum.

PROBLEM 11.1	TEH	1/6

Total weight :

    Piers A & C,

        $W = 2 \times 5{,}405 K / \text{bearing}$

              $= 10{,}810 K$ per Pier

    Pier B,

        $W = 2 \times 12{,}611 K / \text{bearing}$

            $= 25{,}222 K$

    $W_{TOT} = 2 \times 10{,}810 + 25{,}222$

    $\Rightarrow \underline{W_{TOT} = 46{,}842 \text{ kips}}$

Post-yield stiffness, R_d :

    Piers A & C,

        $R_d = \dfrac{W}{R} = \dfrac{5{,}405}{244}$

            $= 22.2 K/in / \text{bearing}$

            $\times 2 \text{ bearings} = 44.4 K/in / \text{pier}$

    Pier B,

        $R_d = \dfrac{12{,}611}{244} = 51.7 K/in / \text{bearing}$

            $\times 2 \text{ bearings} = 103.4 K/in$

    $(R_d)_{TOT} = 2 \times 44.4 + 103.4$

        $\Rightarrow \underline{(R_d)_{TOT} = 192 K/in}$

$$Q_d = \mu W = 0.06 \times 10,810$$
$$= 648.6 \, K, \text{ Piers A \& C}$$

$$Q_d = 0.06 \times 25,222$$
$$= 1,513 \, K, \text{ Pier B}$$

$$(Q_d)_{TOT} = 2 \times 648.6 + 1,513$$
$$\Rightarrow \underline{(Q_d)_{TOT} = 2,810 \text{ kips}}$$

Iteration #1.

Assume $D_{MAX} = D_{ISO} = 18.75''$

$$\frac{D_{MAX}}{R} = \frac{18.75}{244} = 0.077$$

$$k_{EFF} = 192 + 0.06 \left(\frac{46,842}{18.75} \right)$$

$$\Rightarrow k_{EFF} = 342 \, K/in$$

$$T_{EFF} = 2\pi \sqrt{\frac{46,842}{386 \times 342}}$$

$$\Rightarrow T_{EFF} = 3.74 \text{ sec}$$

$$PSA_{TEFF} = 0.20 \, g \quad (\text{geomean})$$

$$\xi_{EFF} = \frac{2}{\pi} \cdot \frac{0.06}{0.06 + 0.077}$$

$$\Rightarrow \xi_{EFF} = 0.279$$

PROBLEM 11.1	TEH	3/6

$$R_\xi = \left(\frac{0.05}{0.279}\right)^{0.3} = 0.597 > 0.588$$

$$\Rightarrow R_\xi = 0.597$$

$$D_{max} = 0.20 \times 386 \times .597 \times \left(\frac{3.74}{2\pi}\right)^2$$

$$\Rightarrow D_{max} = 16.3 \text{ inches}$$

$$\times 1.3 \quad \text{(geomean-to)}$$

$$\underline{\qquad\qquad \text{RotD100}}$$

$$21.2 \text{ inches}$$

$$> (D_{max})_{Assumed} = 18.75''$$

After several iterations, the final is summarized :

Assume $\underline{D_{Max} = 22.97''}$

$k_{EFF} = 314 \text{ K/in}$

$T_{EFF} = 3.90 \text{ sec}$

$P_S A_{T_{EFF}} = .19 \times 1.3 = .247$

$\xi_{EFF} = 0.248$

$R_\xi = 0.619$

$D_{Max} = 22.73''$

$\approx 22.97''$

$\Rightarrow$ converged

PROBLEM 11.1	TEH	4/6

To estimate plan torsional effects, use the method from ASCE 7-16, Chapter 17, Section 17.5.3.3

$$b = 88' \qquad d = 1,800 \, ft$$

$$e = .05 \times 1800 = 90 \, ft$$

$$y = 900 \, ft, \quad \text{Piers A \& C}$$

$$= 0 \, ft, \quad \text{Pier B}$$

$$r = \sqrt{\frac{b^2 + d^2}{12}} \quad \Rightarrow \quad r = 520 \, ft$$

$$P_T = \frac{1}{r}\sqrt{\frac{\sum_1^z (x^2 + y^2)}{N}}$$

$$= \frac{1}{520}\sqrt{\frac{6(44^2) + 2(90)^2 + 2(810)^2 + 2(990)^2}{6}}$$

$$\Rightarrow P_T = 1.424$$

$$D'_{MAX} = D_{MAX}\left[1 + \frac{y}{P_T^2}\left(\frac{12 \times 90}{88^2 + 1800^2}\right)\right]$$

$$= D_{max}(1.148), \quad \text{Piers A \& C}$$

Pier B, $y = 0 \Rightarrow \underline{D_{max} = 22.97''}$

Piers A \& C, $\Rightarrow D_{max} = 22.97 \times 1.148$

$$= \underline{26.37''}$$

| PROBLEM 11.1 | TEH | 5/6 |

$F_{max} = Q_d + k_d D_{max}$

$= 648.6 + 44.4 \times 26.37$

$\Rightarrow \overline{F_{max}} = 1,819 \text{ kips, Piers A & C}$
(total, 2 bearings)

$= 1,513 + 103.4 \times 22.97$

$\Rightarrow \overline{F_{max}} = 3,888 \text{ kips, Pier B}$
(total, 2 bearings)

$D_{vert} = R(1 - \cos\varphi)$ $\varphi = \sin^{-1}(D_{max}/R)$

Piers A & C, $\varphi = 0.1083 \text{ rad}$

$\underline{D_{vert} = 1.43''}$

Pier B, $\varphi = 0.0943 \text{ rad}$

$\underline{D_{vert} = 1.08''}$

PROBLEM 11.1	TEH	6/6

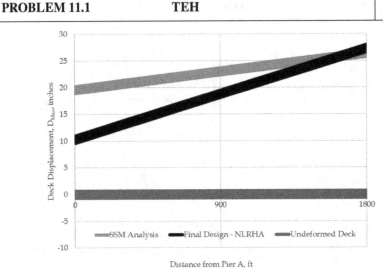

The SSM analysis provides a good sanity check on the final design results from nonlinear response history analysis (NLRHA).

SSM results are slightly higher at the central pier, Pier B:

- 18.75 inches from NLRHA
- 22.97 inches from SSM

SSM results are very close, but slightly lower at the end piers, Piers A and C:

- 27.25 inches from NLRHA
- 26.97 inches from SSM

11.4   EXERCISES

E11.1.

Perform the transverse direction substitute-structure analysis of partially isolated Interstate 40 over State Route 5 (Figures 11.1 and 11.2). A summary of design parameters is given below.

A_S=0.491 g S_{DS}=0.917 g S_{DI}=0.655 g
Steel Beams (15): R_{DL}=275 k/beam R_{LL}=56 k/beam

Isolators:
Q_d=15.8 k/beam k_d=5.7 k/in/beam
Δ_{non-eq}=0.75 inches, non-seismic displacement demand

Superstructure: W_{ss}=29.6 kips/ft, includes six lanes of HL-93 uniform load

Integral abutments: W_{abut}=427 kips, each
                       21 pipe piles per abutment, K_{pile}=20 kips/inch/pile

Spans: 95 ft, 156 ft, 95 ft = 346 ft total bridge length
Piers/Bents: K_{Pier} = 18,106 kips/inch/pier, transverse
                  K_{Pier} = 640 kips/inch/pier, longitudinal
                  ϕV_y = 0.9V_y = 5,171 kips/pier, transverse
                  ϕV_y = 0.9V_y = 1,097 kips/pier, longitudinal
                  V_{po} = 8,359 kips/pier, transverse (overstrength, plastic)
                  V_{po} = 1,748 kips/pier, longitudinal (overstrength, plastic)

W_{HM} = 479 kips/pier, lumped mass for higher mode effects
                  V_u = 101 kips/pier, Strength limit state shear

Bibliography

AASHTO, 2011. *Guide Specifications for LRFD Seismic Bridge Design.* 2nd ed. Shington, D.C.: American Association of State Highway and Transportation Officials.

AASHTO, 2014. *Guide Specifications for Seismic Isolation Design.* 4th ed. Washington, D.C.: American Association of State Highway and Transportation Officials.

AASHTO, 2017. *LRFD Bridge Construction Specifications.* Washington, D.C.: American Association of State Highway and Transportation Officials.

AASHTO, 2020. *LRFD Bridge Design Specifications.* 9th ed. Washington, D.C.: American Association of State Highway and Transportation Officials.

AISC, 2016. *AISC 360-16: Specification for Structural Steel Buildings.* Chicago, IL: American Institute of Steel Construction.

AREMA, 2021. *Manual for Railway Engineering.* Landover, MD: American Railway Engineering and Maintenance-of-Way Association.

Atkinson, G. & Beresnev, I. A., 2002. Ground Motions at Memphis and St. Louis from M 7.5–8.0 Earthquakes in the New Madrid Seismic Zone. *Bulletin of the Seismological Society of America*, 92(3), pp. 1015–1024.

Buckle, I., Constantinou, M., Dicleli, M. & Ghasemi, H., 2006. *Seismic Isolation of Highway Bridges (MCEER-06-SP07)*, Buffalo, NY: Multidisciplinary Center for Earthquake Engineering Research.

Buckle, I. et al., 2006. *Seismic Retrofitting Manual for Highway Bridges (FHWA-HRT-06-032)*, McLean, VA: Federal Highway Administration.

Bunner, M., 2015. *Steel Bridge Design Handbook: Splice Design*, Washington, D.C.: Federal Highway Administration.

City of Los Angeles Harbor Department, 2010. *Code for Seismic Design, Upgrade and Repair of Container Wharves*, The Port of Los Angeles.

Computers and Structures, Inc., 2016. *CSiBridge Version 18.2.0*, Berkeley, CA: s.n.

Fernández, J. A., 2007. *Numerical Simulation of Earthquake Ground Motions in the Upper Mississippi Embayment*, Doctoral Dissertation, Atlanta, GA: Georgia Institute of Technology.

Feygin, V. B., 2015. Performance Based Design of Wharves with Steel Pipe Piles. *Global Journal of Researches in Engineering: Civil and Structural Engineering*, 15(3), pp. 1–16.

FHWA, 2015. *Steel Bridge Design Handbook.* Washington, DC: Federal Highway Administration.

FIP Industriale, 2011. *Lead Rubber Bearings Product Catalog*, Selvazanno, Italy: FIP Industriale.

Fulmer, S. J., Kowalsky, M. J. & Nau, J. M., 2013. *Seismic Performance of Steel Pipe Pile to Cap Beam Moment Resisting Connections*, Raleigh, NC: North Carolina State University.

Giannopoulos, D. & Vamvatsikos, D., 2018. Ground Motion Records for Seismic Performance Assessment: To Rotate ot not to Rotate. *Earthquake Engineering and Structural Dynamics*, 47(12), pp. 2410–2425.

Grubb, M. A., Frank, K. H. & Ocel, J. M., 2018. *Bolted Field Splices for Steel Bridge Flexural Members*, American Institute of Steel Construction / National Steel Bridge Alliance.

Gulkan, P. & Sozen, M. A., 1974. Inelastic Response of Reinforced Concrete Structures to Earthquake Ground Motions. *Journal of the American Concrete Institute*, 71(12), pp. 604–610.

Hadjian, A. H., 1981. On the Correlation of the Components of Strong Motion - Part 2. *Bulletin of the Seismological Society of America*, 71(4), pp. 1323–1331.

Hajihashemi, A., Pezeshk, S. & Huff, T., 2017. A Comparison of Nonlinear Static Procedures and Modeling Assumptions for Seismic Design of Ordinary Bridges. *ASCE Practice Periodical on Structural Design and Construction*, 22(2), pp. 1–10.

Hallmark, R., 2006. *Low Cycle Fatigue of Steel Piles in Integral Abutment Bridges*, Lulea, Sweden: Lulea University of Science anf Technology.

Harn, R., Ospina, C. E. & Pachakis, D., 2019. *Proposed Pipe Pile Strain Limits for ASCE 61-19*. American Society of Civil Engineers, pp. 437–448.

Hashash, Y. M. A., Tsai, C.-C., Phillips, C. & Park, D., 2008. Soil-Column Depth-Dependent Seismic Site Coefficeints and Hazard Maps for the Upper Mississippi Embayment. *Bulletin of the Seismological Society of America*, 98(4), pp. 2004–2021.

Helwig, T. & Yura, J., 2015. *Steel Bridge Design Handbook: Bracing System Design*, Washington, D.C.: Federal Highway Administration.

Huff, T., 2014. Spanning the Wolf River Wetlands. *ASPIRE - The Concrete Magazine*, pp. 14–17.

Huff, T., 2016a. Partial Isolation as a Seismic Design Strategy for Pile Bent Bridges in the New Madrid Seismic Zone. *ASCE Practice Periodical on Structural Design and Construction*, 21(2), pp. 1–12.

Huff, T., 2016b. Seismic Displacement Estimates for Bridges in the New Madrid Seismic Zone. *ASCE Practice Periodical on Structural Design and Construction*, May, 21(2), pp. 1–9.

Huff, T., 2016c. Structural Demand on Bridges Subjected to Bidirectional Ground Motions. *ASCE Practice Periodical on Structural Design and Construction,* 22(1), pp. 1–13.

Huff, T. & Pezeshk, S., 2016. Inelastic Displacement Spectra for Bridges Using the Substitute Structure Method. *ASCE Practice Periodical on Structural Design and Construction*, 21(2), pp. 1–13.

Huff, T. & Shoulders, J., 2017. *Partial Isolation of a Bridge on Interstate 40 in the New Madrid Seismic Zone*, National Harbor, MD: Engineers Society of Western Pennsylvania.

Kawashima, K., 2004. Seismic Isolation of Highway Bridges. *Journal of Japan Association for Earthquake Engineering*, 4(3, Special Issue), pp. 283–297.

Kottke, A. & Rathje, E. M., 2008. A Semi-Automated Procedure for Selecting and Scaling Recorded Earthquake Motions for Dynamic Analysis. *Earthquake Spectra*, 24(4), pp. 911–932.

Kramer, S. L., Arduino, P. & Sideras, S. S., 2012. *Earthquake Ground Motion Selection*, Seattle, WA: Washington State Department of Transportation.

Lehman, D. & Roeder, C. W., 2012. *An Initial Study into the use of Concrete Filled Steel Tubes for Bridge Piers and Foundation Connections*, Seattle, WA: The university of Washington.

Malekmohammadi, M. & Pezeshk, S., 2014. Nonlinear Site Amplification Factors for Sites Located within the Mississippi Embayment with Consideration for Deep Soil Deposit. *Earthquake Spectra*.

Maurer, 2011. *Seismic Isolation Systems with Lead Rubber Bearings*, Munich, Germany: Maurer Sohne.

Miller, R. A., Castrodale, R., Mirmiran, A. & Hastak, M., 2004. *NCHRP Report 519: Connection of Simple-Span Precast Concrete Girders for Continuity*, Washington, D.C.: National Academies of Science, Engineering and Medicine.

Moon, S.-w., Hashash, Y. M. A. & Park, D., 2016. USGS Hazard Map Compatible Depth-Dependent Seismic Site Coefficients for the Upper Mississippi Embayment. *KSCE Journal of Civil Engineering*, 21, pp. 1220–1231.

National Academies of Sciences, Engineering, and Medicine, 2021. *Proposed Modification to AASHTO Cross-Frame Analysis and Design*, Washington, D.C.: The National Academies Press.

NEHRP Consultants Joint Venture, 2011. *Selecting and Scaling Earthquake Ground Motions: NIST GCR 11-917-15*, Redwood City, California: National Institute of Standards and Technology.

Pacific Earthquake Engineering Research Center, 2014. *PEER Ground Motion Database*. [Online]; Available at: http://ngawest2.berkeley.edu/ [Accessed 5 May 2017].

PCI, 2014. *Bridge Design Manual*, Chicago, IL: Precast Prestressed Concrete Institute.

PEER, 2010. *Technical Report for the PEER Ground Motion Database Web Application*. Berkeley, CA: Pacific Earthquake Engineering Research Center.

Pietra, G. M., Calvi, G. M. & Pinho, R., 2008. *Displacement-Based Seismic Design of Isolated Bridges, Research Report ROSE - 2008/01*. 1st ed. Pavia, Italy: IUSS Press.

POLA, 2010. *The Port of Los Angeles Code for Seismic Design, Upgrade and Repair of Container Wharves*, Los Angeles, CA: Port of Los Angeles.

POLB, 2015. *Port of Long Beach Wharf Design Criteria*, Long Beach, CA: Port of Long Beach.

Port of Long Beach, 2012. *Wharf Design Criteria,* s.l.: s.n.

Priestley, M. J. N., Calvi, G. M. & Kowalsky, M. J., 2007. *Displacement-Based Seismic Design of Structures*. 1st ed. Pavia, Italy: IUSS Press.

Priestley, M. J. N. & Grant, D. N., 2005. Viscous Damping in Seismic Design and Analysis. *Journal of Earthquake Engineering - Imperial College Press*, 9, pp. 229–255.

Priestley, M. J. N., Seible, F. & Calvi, G. M., 1996. *Seismic Design and Retrofit of Bridges*. 1st ed. New York, NY: John Wiley & Sons.

Robinson Seismic LTD, 2011. *Catalog of Robinson Seismic Bearings*, Petone, New Zealand: Robinson Seismic LTD.

Rollins, K. M. & Stenlund, T. E., 2008. *Laterally Loaded Pile Cap Connections*, Provo, Utah: Brigham Young University.

Romero, S. M. & Rix, G. J., 2005. *Ground Motion Amplification of Soils in the Upper Mississippi Embayment*, Urbana, IL: NSF/MAE Center.

SeismoSoft, 2020a. *SeismoArtif 2020*. [Online]; Available at: http://seismosoft.com/en/SeismoArtif.aspx

SeismoSoft, 2020b. *SeismoMatch 2020*. [Online]; Available at: http://seismosoft.com/en/SeismoMatch.aspx

Silva, P. F. & Seible, F., 1999. *Design Example of a Multiple Column Bridge Bent Under Seismic Loads Using the Alaska Cast-in-Place Steel Shell*, San Diego, CA: University of California.

Silva, P. F. & Sritharan, S., 2011. Seismic Performance of a Concrete Bridge Bent Consisting of Three Steel Shell Columns. *Earthquake Spectra*, 27(1), pp. 107–132.

Stafford, P. J., Mendis, R. & Bommer, J. J., 2008. Dependence of Damping Correction Factors for Response Spectra on Duration and Number of Cycles. *ASCE Journal of Structural Engineering*, 134(8), pp. 1364–1373.

Stenlund, T., 2007. *Laterally Loaded Pile Cap Connections*, Provo, UT: Brigham Young University.

Stephens, J. & McKittrick, L., 2005. *Performance of Steel Pipe Pile-to-Concrete Bent Cap Connections Subject to Seismic or High Transverse Loading: Phase II*, Bozeman: Montana State University.

Stephens, M. T., Berg, L. M., Lehman, D. E. & Roeder, C. W., 2016. Seismic CFST Column-to-Precast Cap Beam Connections for Accelerated Bridge Construction. *ASCE Journal of Structural Engineering*, 142(9), pp. 1–14.

Stevens, J. E. et al., 1998. *Performance of Steel Pipe Pile-to-Concrete Bent Cap Connections Subject to Seismic or High Transverse Loading: Phase I*, Bozeman: Montana State University.

Tsopelas, P., Nagarajaiah, S., Constantinou, M. C. & Reinhorn, A. M., 1991. *3D-BASIS-M: Nonlinear Dynamic Analysis of Multiple Building Base Isolated Structures*. Technical Report NCEER-91-0014. Buffalo, NY: University of Buffalo.

United States Geological Survey, 2020. *USGS Unified Hazard Tool*. [Online]; Available at: https://earthquake.usgs.gov/hazards/interactive/ [Accessed 6 January 2020].

USGS, CGS, ANSS, 2014. Center for Engineering Strong Motion Data. [Online]; Available at: http://strongmotioncenter.org/

Warn, G. P., 2002. Displacement Estimates in Isolated Bridges. *MCEER Student Research Accomplishments*.

Warn, G. P. & Whittaker, A. S., 2007. *Performance Estimates for Seismically Isolated Bridges (MCEER-07-0024)*, Buffalo, NY: Multidisciplinary Center for Earthquake Engineering Research.

Watson-Lamprey, J. A. & Boore, D. M., October 2007. Beyond SA-GMRotI: Conversion to SA-Arb, SA-SN, and SA-MaxRot. *Bulletin of the Seismological Society of America*, 97(5), pp. 1511–1524.

Wen, Y. K., 1976. Method for Random Vibration of Hysteretic Systems. *Journal of the Engineering Mechanics Division*, 102, pp. 246–263.

WSDOT, 2020. *Bridge Design Manual*, Olympia, WA: Washington State Department of Transportation.

Yura, J. A., 2001. Fundamentals of Beam Bracing. *Engineering Journal*, 38, pp. 11–26.

Index

Printed in the United States
by Baker & Taylor Publisher Services